Psychological Operations
Principles and Case Studies

Editor

Frank L. Goldstein, Col, USAF

Co-editor

Benjamin F. Findley, Jr., Col, USAFR

Air University Press
Maxwell Air Force Base, Alabama

September 1996

Library of Congress Cataloging-in-Publication Data

Psychological operations : principles and case studies / editor, Frank L. Goldstein ;
    co-editor, Benjamin F. Findley.
      p. cm.
    At head of t.p. : AU Shield.
    "September 1996."
    1. Psychological warfare—United States. 2. Psychological warfare—Case studies.
      I. Goldstein, Frank L., 1945– . II. Findley, Benjamin F.
      UB276.P82 1996
      355.3'434—dc20

                                              96-22817
                                              CIP

ISBN 1-58566-016-7

First Printing September 1996
Second Printing January 1999
Third Printing June 2000
Fourth Printing September 2002
Fifth Printing June 2003
Sixth Printing November 2003

Disclaimer

For Sale by the Superintendent of Documents
US Government Printing Office
Washington, DC 20402

Contents

PART I

Nature and Scope of
Psychological Operations (PSYOP)

PART IV

Case Studies of PSYOP Applications

Illustrations

Foreword

From a military commander's perspective, the role of psychological operations (PSYOP) in the successful planning and execution of modern military operations is absolutely essential. The recent successes of PSYOP in Panama and the Persian Gulf amply demonstrate its importance in achieving our military objectives and national goals. PSYOP is an integral part of the United States Special Operations Command mission. Today, the military faces a dynamic and unpredictable world. PSYOP will remain a valuable instrument in our overall defense posture and will be a key asset in the fulfillment of US national policy.

Psychological Operations: Principles and Case Studies serves as a fundamental guide to PSYOP philosophy, concepts, principles, issues, and thought for both those new to, and those experienced in, the PSYOP field and PSYOP applications. This book clarifies the value of PSYOP as a cost-effective weapon and incorporates it as a psychological instrument of US military and political power, especially given our present budgetary constraints. The authors contribute to the understanding of psychological operations by presenting diverse articles that portray the value of the planned use of human actions to influence perceptions, public opinion, attitudes, and behaviors so that PSYOP victories can be achieved in war and in peace. The four sections classify articles with related themes into a common category.

Part I offers an overview of the nature and scope of PSYOP and serves as an introduction to the overall nature, historical background, concepts, and principles of psychological operations. These independent articles, which reflect the broad scope of historical development and thought about PSYOP, are intended to be a foundation for understanding the basic nature and key elements of PSYOP.

Part II follows with issues and influences related to developing effective US strategy, doctrine, and structure for conducting psychological and political warfare. The authors focus on those psychological issues and roles that have been recurrent as our national policy, objectives, and strategy have been formulated and implemented. They explain historical and contemporary elements of the national policy process and the framework within which national PSYOP policy is formulated, administered, and implemented.

Part III deals with the objectives and activities of strategic, tactical, operational, and other types of PSYOP. These authors emphasize that all forms of PSYOP should primarily support the attainment of national policy and objectives. They conclude that the key to all US PSYOP is credibility of the message as defined by the influencing or changing of perceptions, attitudes, and behaviors through utilization of words and actions.

In Part IV the authors use case studies to present and clarify PSYOP goals, roles, and methods. One of the editors of this book condenses and analyzes (1) US and Vietcong PSYOP in the Vietnam War and (2) the Iraqi propaganda network. The other writers examine tactical and consolidation PSYOP activities in Operations *Just Cause* and *Promote Liberty* in Panama. They address (1) our national antidrug policy and its relationship to the role of military psychological operations and (2) the importance of the political-psychological dimensions of conflict and insurgency.

By addressing the breadth and depth of psychological operations thought, this collection of PSYOP articles serves as a valuable knowledge base for those who read it. A major purpose of the book is to pull together those previously published articles under one cover in one volume. *Psychological Operations* should stimulate your thinking and reinforce the value of PSYOP.

Frank L. Goldstein, PhD
Col, USAF
Dean of Research
Air Command and Staff College

Preface

This book explores the breadth and depth of fundamental PSYOP roles, principles, and methods. Many of these readings were previously published in professional journals. *Psychological Operations: Principles and Case Studies* is not a collective work and does not attempt to be. Each article is an independent effort and together they represent a cross section of what the "best and brightest" feel is key to both offensive and defensive psychological operations. What the 24 contributors share, besides their own particular expertise in PSYOP, is a recognition that a clear understanding of PSYOP is important to our national defense and to the way PSYOP should be conducted in the future.

Understanding PSYOP is not a simple task. Historically, both military and civilian discussions of PSYOP throughout the leadership spectrum have regularly substituted clichés, myths, and untruths for hard evidence or analysis of what PSYOP is and how it can serve our national objectives. PSYOP policy and doctrine have not received their deserved attention while hostile PSYOP efforts against the US are misunderstood and often ineffectively countered.

Although the Soviet threat has ended, three Soviet PSYOP articles are deliberately included because the classic Soviet PSYOP model has been taught and integrated into third world countries all over the world and will continue to be an influence in the international arena.

In 1984, a major effort began to revitalize the state of psychological operations in the United States. That effort was spearheaded in 1985 by a document known as the DOD PSYOP Master Plan. The plan evolved over several years and revision began again in 1990, just prior to the Persian Gulf War. The plan directed the overall revitalization of US psychological operations; it directed that PSYOP be fully integrated into planning and military operations; it directed that PSYOP be considered for immediate and effective use in crises and hostilities; and it directed the use of overt PSYOP programs in peacetime. The plan presented an across-the-board baseline of problem areas and laid out specific remedial actions for fundamental improvements in our capability to employ PSYOP in peace, crisis, and war.

This volume is also an ongoing effort. While the collected articles will expand both the knowledge and understanding of psychological operations past and present, many of these authors would not have written their articles without the initial revitalization—and Air University Press may not have published them had they been written. The essays present critical PSYOP issues, problems, activities, and techniques in the book's four sections, which classify articles with related themes into a common category.

We must underscore the fact that these essays are independent of each other. Many were written and published decades ago, and most were originally published in differing formats with differing style requirements. Those that were written in the present tense years ago have been edited to reflect the past tense. We have retained the original flavor of each essay, but some style changes have been made to conform more closely to current Air University Press style. Some essays have notes, some don't; some have bibliographies, some don't; some have appendixes, some don't. We have retained the original styles, with this exception: Notes number 1 that appeared at the titles of some original essays have been removed and the references thereto have been moved to the bottom of the first page.

We are grateful to the 24 contributing authors and regret that we could not include biographical information on all of them; unfortunately, we were unable to obtain biographical data on Preston S. Abbott. We thank J. Schmier, B. Karabaich, and C. Williamson, who provided assistance to MSgt Blair and Col Goldstein, and we are especially grateful to Maj Thomas P. Sands, Florida ANG, who provided editorial assistance to the entire project in its early stages. We also extend our gratitude to those who provided outstanding administrative and technical support. In particular, Majors Stephen Asher, USAF, and James V. Keifer, USAF, Retired; Ms Victoria L. Blessing and Ms Eva E. Hensley; SFC Roger D. Crocker, USAR; SSgt Sherri Adler, USAFR; Sgt Manuel D. Gonzalez, USAFR; and A1C Jayme L. Laurent, Florida ANG.

PART I

Nature and Scope of Psychological Operations (PSYOP)

Introduction

The scope of military PSYOP during World War II, the Korean War, and in much of the 1960s was primarily limited to combat propaganda and psychological warfare (psywar). During those times, it was accepted as a specialized tactical application and as a subordinate operation. The experiences of these conflicts, especially the Vietnam War, convinced some American military and political leaders that the psychological dimension of national power and conflict encompasses diverse elements and many activities—nonmilitary as well as military—in both peacetime and war in support of our national policy and objectives. Its scope can vary from the tactical battlefield to the operational theater to the strategic levels of conflict to national political and military goals.

Part I serves as an introduction to the overall nature, historical background, and concepts of PSYOP, and to some principles that can be used for training in the field of psychological operations. The independent articles in this section reflect the broad range of historical development and thought about PSYOP and are intended to be a foundation for understanding the basic nature and key elements of PSYOP.

Col Frank L. Goldstein, USAF, and Col Daniel W. Jacobowitz, USAF, Retired, provide a general introduction to and a commonly accepted definition of PSYOP. The authors explore the three types of PSYOP and give several examples of strategic, tactical, operational, and consolidation PSYOP. They divide propaganda into white, gray, and black classes, and present the various resources of psychological operations. The six major military objectives of PSYOP are condensed for the reader.

The late Col Fred W. Walker, USAF, Retired, presents strategic concepts for military PSYOP to enhance the overall understanding of the PSYOP dimension and its challenges. His emphasis is on the public's understanding that the US government does *not* engage in public disinformation activities in pursuit of national policy. He addresses strategic rather than tactical battlefield and operational PSYOP. He focuses on the concept that "truth is the best propaganda" and analyzes efforts to counter Soviet global disinformation. Walker died before the dissolution of the Soviet Union; however, the article represents a clear and precise use of Soviet PSYOP between 1917 and 1990.

Col Alfred H. Paddock, Jr., USAF, Retired, provides a brief perspective on the evolution of military PSYOP. He addresses the 1985 Department of Defense (DOD) PSYOP Master Plan as the major framework for rebuilding worldwide military PSYOP capabilities in support of national objectives in peace and crisis and at all levels of conflict. Colonel Paddock presents several essential themes for remedial actions—developing comprehensive joint PSYOP doctrine, developing an adequate number of PSYOP planners on joint staffs, and improving commanders' understanding of PSYOP capabilities and missions.

Col Benjamin F. Findley, Jr., USAFR, analyzes the many similarities between civilian business marketing and military PSYOP. His premise is that, just as marketing and promotion processes in business have dedicated necessary resources and utilized persuasive strategies and tactics to successfully influence perceptions, motivations, attitudes, and opinions to achieve results, so military PSYOP can and should employ similar approaches to market our country. Colonel Findley focuses on basic persuasion and marketing principles and on business lessons learned.

Psychological Operations

An Introduction

Col Frank L. Goldstein, USAF
Col Daniel W. Jacobowitz, USAF, Retired

> *If your opponent is of choleric temper, try to irritate him. If he is arrogant, try to encourage his egotism. If the enemy troops are well prepared after reorganization, try to wear them down. If they are united, try to sow dissension among them.*
>
> — General Tao Hanzhang; translated by Yuan Shibing
> *Sun Tzu's Art of War: The Modern Chinese Interpretation*

PSYOP is a vital element within the broad range of US political, military, economic, and ideological actions. Properly employed, PSYOP reduces the morale and combat efficiency of enemy troops and creates dissidence and disaffection within their ranks. Psychological operations can promote resistance within a civilian populace against a hostile regime or be employed to enhance the image of a legitimate government. The ultimate objective of American PSYOP is to convince enemy, friendly, and neutral nations and forces to take action favorable to the US and its allies. Because of the nature of the parent society and the comparative case of detecting falsehood in a multimedia world, US overt PSYOP campaigns are limited to presenting factually correct material. It would be disingenuous to claim that a balanced picture is presented in US propaganda, but the actual material presented in any particular overt PSYOP message will be verifiable against independent sources.

Truth and falsehood in propaganda must be separated from overt and covert operations and the issue of white, gray, and black (false) propaganda. Overt propaganda is produced by a government or organization that takes responsibility for it. Because of police state conditions or tactical considerations, it

5

may have to be disseminated by covert means, such as agents who risk their lives to transport and distribute the materials. Overt propaganda may be true or false. Since the effect of propaganda depends on credibility, overt sources that utilize falsehoods quickly lose all effectiveness. Overt propaganda is also known as white propaganda because the source takes responsibility for it. Gray propaganda is material that is distributed without an identified source. It may be true or false. Black propaganda is material produced by one source that purports to have emanated from another source. Such covert productions may be used to damage the credibility of a white (truthful) source by disseminating obvious falsehoods under the label of the previously trusted source. Black propaganda—if effective at all—quickly loses effectiveness unless the populace is particularly susceptible to rumors, manipulation, and distortion of fact. Nevertheless, black propaganda can be highly effective if properly planned. For example, should intel sources determine that an invasion is imminent, broadcasting that fact under an aegis purporting to be that of the potential invader removes all surprise and falsifies all the invader's claims of a "just" war.

Propaganda may legitimately be economical of the truth. For example, in describing the triumph of democracy there is no particular obligation to discuss the role of Boss Tweed in urban politics.

Military psychological operations are inherently joint operations. Unified, joint task force, and other military commanders identify target audiences and develop PSYOP themes, campaigns, and products. These are submitted through channels to the joint chiefs for approval. The principles of developing a PSYOP campaign are applicable across the operational continuum. Although the complexity of the methodology varies with the level of conflict, considerations for development of PSYOP campaigns are the same for counterterrorism as they are for global war.

The psychological dimension covers the battlefield as well as the effects upon the soldiers fighting the battle, their military leaders and staffs, the political leaders, and the civilian population. On the field of battle, US forces want to face an enemy who is both unsure about his cause and capabilities and sure about his impending doom; an enemy who, even if unwilling to surrender, has little will to engage in combat.

It is US policy that psychological operations will be conducted across the operational continuum. It must be understood that psychological operations are conducted continuously to influence foreign perceptions and attitudes in order to effect changes in foreign behavior favorable to US national security objectives. Any type or level of PSYOP can be conducted at any point along the operational continuum. The operational environment in which psychological operations are conducted does not, by itself, dictate or limit PSYOP actions or the level of PSYOP applied.

In environments short of declared war, national PSYOP policy is normally derived from official policy statements and declarations on US foreign policy as well as national security policy. Interagency coordination is required. During declared war, the policy emanates from the national command authorities (NCA) upon approval of plans submitted by the Office of the Secretary of Defense (OSD). This national policy is executed through a strategy of coherent international information programs, which consist of US information dissemination efforts dealing with policy and information. It is essential that PSYOP themes and products reflect and support national policy; these overt messages are as official as any White House press release. Therefore, appropriate PSYOP policy and strategy must fully integrate Department of Defense (DOD) PSYOP into these international information programs to alleviate the potential for disseminating contradictory information.

Psychological actions such as show of force, cover, and deception have been used throughout history to influence enemy groups and leaders. Modern psychological operations are enhanced by the expansion of mass communication capabilities. Nations can multiply the effects of their military capabilities by communicating directly to their enemies a threat of force or retaliation, conditions of surrender, safe passage for defectors, incitations to sabotage, support to resistance groups, and other messages. The effectiveness of this communication on the target audience depends on their perceptions of the communicator's credibility—does the communicator have the capability to carry out the threatened actions?

PSYOP actions convey information not only to intended target audiences but also to foreign intelligence systems. Therefore, PSYOP messages must be coordinated with cover and deception plans and activities, along with operational security planners, to ensure that essential secrecy is realized and that PSYOP messages reinforce cover and deception objectives. Skillful content analysts can determine overall intentions by carefully analyzing PSYOP messages and PSYOP planners can screen their own products to ensure that only the overt intention is broadcast. The methodology of overt propaganda analysis is arcane and difficult, as much derived from art as science. Some practitioners believe the method is more valid when aimed at totalitarian propaganda than PSYOP produced by democracies. Democratic propaganda normally is far less patterned, possibly because the products reflect a less organized process—ad hoc arrangements, swiftly evolving policies, lack of hidden agendas, or, frequently, no agendas at all. Totalitarian—especially communist—propaganda may be easier to analyze because it is highly formalized and patterned.

There is a psychological dimension within any element of national power projection, particularly the military element. Foreign perceptions of US military capabilities are fundamental to strategic deterrent capability. Therefore, US policymakers must articulate our national and military actions (if we don't, others will). Communicating unambiguously to allies, enemies, and neutrals is a key element of US national strategy. The effectiveness of deterrence, power projection, and other strategic concepts hinges on our ability to influence the perceptions of others.

For these communications, any player in the US government or overall body politic may become an important tactical element regardless of the strategic position that the player holds. In conveying the will of the United States, the firm set of the president's jaw in drawing "a line in the sand" may have as much influence on international, and especially adversarial, understanding of US policy as the actions taken by the government. Supporting statements by other officials, including the secretary of state, congressional leaders, and military commanders, similarly are tactical elements carrying out the information strategy. Tactical actions of this nature, delivered at the strategic level, are analogous to the actual tactical delivery of weapons to targets of a strategic nature in a shooting war. Since much of policy is devoted to achieving national goals while ameliorating genuine conflict and avoiding a shooting war, tactical performance of these roles by strategic elements of the political/military system is critical to national policy. One of the benefits of the open political process—disseminated and monitored by a free and aggressive media—is that individuals who would be inadequate tactical communicators tend to be shunted away from positions for which the nation's fate requires skillful performance. Military PSYOP may be undertaken at the strategic level, augmenting other national communications systems, particularly in areas for which the peacetime national systems—such as the United States Information Agency—have no access.

In every case, it is crucial that military PSYOP be integrated with other national communications, since the audiences will accept military PSYOP messages as official positions. To ensure this process, military psychological operations rely on a planned, systematic process of conveying messages to, and influencing, selected foreign groups. The messages conveyed by military PSYOP are intended to promote particular themes that result in desired foreign attitudes and behaviors. Therefore, PSYOP may be used to establish and reinforce foreign perceptions of US military capability, determination, and responsiveness to US political goals and to support overall US policy.

Psychological operations are an important dimension of overall military operations. They may be used by commanders to influence the attitudes and behavior of foreign groups in a manner favorable to the achievement of US national objectives. Thus, the principal purpose of DOD PSYOP is to persuade foreign audiences to change or enhance attitudes or behaviors in a manner favorable to one or more national security objectives. Additionally, PSYOP can counter foreign propaganda that adversely affects the achievement of US objectives.

The United States typically distinguishes between PSYOP on a strategic level and PSYOP on a tactical, battlefield level.

Strategic psychological operations are usually considered an aspect of public diplomacy and are normally established and guided by intergovernment working groups created for a particular short-term situation or regional area of concern. The intergovernment groups meet periodically to clarify strategic PSYOP policy in light of political and military developments of the day. At the present time, however, the US government has no permanent mechanism to institutionalize this process.

In tactical or battlefield PSYOP, commanders use such techniques as loudspeaker broadcasts and leaflet drops with the intent of generating a force multiplier without having to increase force size. Psyopsers support tactical deception, counterterrorism, counterpropaganda, and other nontraditional means as the tactical situation merits. PSYOP messages cannot replace tactical performance or redeem inadequate training, weapons, or tactics that result in poor combat performance. However, the methodology can increase the overall functional degradation of enemy capability. Missiles, bombs, bullets, and maneuvers establish the context for PSYOP multiplication and hastening the cumulative results of tactical competence. Psychological operations speed the positive effects of military prowess, and may, under certain conditions, delay the consequences of military failure. Because psychological operations multiply desired effects, positive outcomes can result in quicker victory at lower cost in material, time, and casualties. Whether strategic or tactical, PSYOP uses any available means of communication to achieve desired ends. In Western circles, truthfulness is a desirable goal in itself, and is the principal means for building credibility among targeted audiences. Success in PSYOP rests on thorough analysis and planning.

Modern PSYOP planning includes a target analysis that consists of several phases. The first phase identifies possible target audiences. Once the target audience is identified, such target characteristics as vulnerabilities, susceptibilities, conditions, and effectiveness are analyzed. Vulnerabilities are the four psychological factors that affect the target audience: perception, motivation, stress, and attitude. Susceptibilities include the degree to which the target audience can be influenced to respond to the message it receives. Conditions of the target audience include all environmental factors—social, economic, political, military, and physical—that influence the target audience. Audience effectiveness is the capability of the target audience to carry out the psyopser's desired response. The concept of audience effectiveness is fundamental to PSYOP success at strategic and tactical levels. If the goal is functional destruction of an enemy tactical unit, the effective audience may be individual soldiers, who may be persuaded to desert, defect, or defect in place; that is, simply fail to perform without overtly resisting their commanders. Other goals may require finding different effective audiences. The responsive audience in a battlefield air interdiction campaign could be the civilian workers who repair damaged railroads and bridges. Truthfully reporting that they are at risk from restrikes of previously damaged targets may dissuade them from voluntarily working. However, if they are slave laborers, the audience may be nonresponsive regardless of their susceptibility. The responsive audience may be taskmasters, or high-level commanders. For example, the susceptibility of the high-level audiences may be threats of war crimes prosecution. Both audiences will have to be convinced, by multiple messages, if the campaign is to be effective. Once the above analysis of audiences is accomplished, the psyopser seeks to determine the specific psychological plan that supports the national objective.

Psychological operations have been a part of military strategy since armies first took the field of battle. The Persian Gulf War and the employment of PSYOP by both sides were the most recent chapter in a long history of PSYOP as an integral part of military strategy. Throughout much of military history, PSYOP's presence has been felt in battlefield campaigns. Psychological operations were integrated into the commander's scheme of maneuver before the label of PSYOP was invented and without the benefit of thorough or scientific planning. An early example of how PSYOP was planned and applied in ancient battle is contained in the writings of the Chinese strategist Sun Tzu, who stated that the most noble victory was to subdue his enemy without a fight. Another was the successful exploits of Genghis Khan (the Mongolian general Temujin), who would soften his enemy's will to resist by spreading rumors about his own army's strength and fierceness. His planning was simple and, seemingly, relevant and effective.

As early as the Battle of Bunker Hill, colonial military PSYOP operators used leaflets designed to work on the susceptibilities of the effective audience. Leaflets distributed among British troops in Boston by trusted colonial agents were based upon analysis of the situation and the conditions the British troops anticipated, as well as their motivation. Thus, some leaflets reported that food and provisions among the colonial troops were far superior to the hardtack fed the British, and that switching sides would result in an immediate improvement in diet. More important and effective was an appeal to a basic desire to improve the British soldier's status in life, an important factor in motivating enlistment. Many troops had joined merely to obtain subsistence or with the hope of achieving enough riches to obtain farmland. The most effective message slightly pointed out that to obtain land in the colonies, a soldier needed merely to desert and walk west until he found a suitable plot. Hessian mercenaries in particular responded to this appeal later in the war, and a considerable number of the present-day Pennsylvania Dutch owe their ancestry to the effectiveness of this appeal as these soldiers settled in a language-compatible area in which they were unlikely to be turned over to British authority. Colonial strategic psychological operations were masterful from inception, with Thomas Jefferson and Thomas Paine effectively working their various chosen audiences while Benjamin Franklin used his post in France to bolster not only continental support—which eventually resulted in the Franco-American force that was victorious at Yorktown—but also helped bring Lafayette, Pulaski, and Kosciuszko to American shores. The campaigns within Britain that drained political support for the war were most effective. Battlefield competence was important to the success of this effort and included not only victories at Trenton and other places but also the amazing raids by John Paul Jones upon English coastal towns, whose political effect far outshone their minimal military importance. James H. Doolittle's raid on Japan—undertaken for the same purposes and with analogous military results—was foreshadowed in methodology and in equivalent technological means almost 150 years before.

In the American Civil War, both sides of the conflict directed strategic campaigns at England in the hope of winning support for their respective causes. It remains questionable, however, that these campaigns had been formally planned and that the proper resources were marshaled to execute them. The Southern campaign was virtually undercut by the Confederate refusal to sell cotton to Great Britain. This economic suicide overwhelmed any positive effects that media campaigns may have engendered.

During World War I, PSYOP came into its own as a formal activity. Almost all countries involved in the war used forms of strategic and tactical PSYOP. Many countries formed military units specializing in propaganda. These units' primary duties included distribution of leaflets by balloon and aircraft. The linkage among planning, resource mobilization, and execution by these agencies appeared to be an uncomplicated matter. How the PSYOP details were integrated into the shooting war of the day, or how well PSYOP induced surrenders, was not recorded for history. What is known, however, is that surrenders occurred with a positive correlation to PSYOP activities. Thus, military analysts began taking a new look at PSYOP as an ingredient with surprising impact on the battle. Psychological operations were a resource because they induced stress on both civilian and military forces of the enemy.

During World War II, propaganda activities became known as psychological warfare (psywar). Public broadcast radio, about 20 years old at this point, was called into play. Tank-mounted loudspeakers with a range of approximately two miles amplified the ability of the human voice to reach opposing combatants. Besides media programs, military actions were undertaken for their PSYOP effect. The Doolittle raid against Japan was considered an important PSYOP event for at least two reasons. The carefully planned raid demonstrated credibly to the Japanese that the US could reach and bomb their homeland, prompting them to take unnecessary steps for home defense. More important perhaps, news of the success back home caused morale to soar in an American population desperate for a victory. Planning, mobilization, and execution all worked in this one instance. However, it must be noted that during this war, aircrews frequently expressed reluctance to risk themselves on leaflet-dropping missions because they lacked confidence in that methodology as a means of bringing victory nearer.

In the years that followed, PSYOP matured as a combat force multiplier, albeit through a series of starts and stops. During the 1950s, the Soviet Union made great strides in both strategic and internal PSYOP. Soviet client-states began very elaborate psychological operations for foreign insurgents and home consumption. At the same time, little was apparently being planned in Western PSYOP circles.

Although strategic and tactical psychological operations were effectively integrated by the North Vietnamese during the Vietnam era, US PSYOP planning was not effectively formalized or coordinated with operations and troop mobilizations. It was in Vietnam that propaganda activities assumed the current term *PSYOP*—and television was a new medium. The North Vietnamese mastered the art of using the international media, particularly television, for their PSYOP. The US government was ineffective in both public information and public policy in mobilizing its public for the war. As a result of this negative experience in Vietnam, the US government learned the importance of domestic and foreign support of major policy goals.

In the more recent conflicts, PSYOP has been integrated with combat operations. In the Falklands, Afghanistan, Africa, South and Central America, Grenada, Panama, and the Persian Gulf, PSYOP was included by all parties. PSYOP even became a critical part of the terrorist mode of operations during the seventies and were part of the Iraqi PSYOP plan when they threatened terrorist activities.

Any student of PSYOP will quickly learn how important PSYOP can be in political and military strategy. What every student should strive for is an internalization of the concept proposed by Sun Tzu, the Chinese military strategist, that to fight and conquer in all your battles is not supreme excellence; supreme excellence consists of breaking the enemy's resistance without resorting to fighting. Because soldiers and civilians have not fundamentally changed in nature or psychology since Sun Tzu wrote these observations, they remain appropriate today.

Strategic Concepts for Military Operations

Col Fred W. Walker, USAF, Retired*

For many years, people have indicated a lack of under-standing of exactly what military psychological operations (PSYOP) entail. News media refer to PSYOP as psychological warfare (psywar) and imply that there is some nefarious objective or purpose in such action. They usually indicate that there is some element of deliberate misinformation—or even some public lies—involved in such activity. It is time that these misperceptions were set straight. The public must understand that the United States government does *not* engage in public disinformation activities in pursuing national policy. This article concentrates on peacetime strategic concepts, setting aside the loudspeaker and leaflet activities of the battlefield.

Joint Pub 1-02, *Department of Defense Dictionary of Military and Associated Terms* (1 December 1989) broadly defines strategic psychological activities as "planned psychological activities in peace and war which normally pursue objectives to gain the support and cooperation of friendly and neutral countries and to reduce the will and the capacity of hostile or potentially hostile countries to wage war."

This is demonstrably any program that supports a long-term effort to achieve a national or regional foreign policy objective through persuasion. We might consider the term *persuasive communications* to mean the same thing as psychological operations.

How did we become focused on strategic psychological activities? Quite fundamentally, various administrations assessed that the United States was falling behind in world

---

*Deceased 1990.

17

influence because the global propaganda and disinformation efforts of the Soviet Union were directed against US foreign policy objectives. As a result of this assessment, President Ronald Reagan directed the Department of Defense (DOD) to revitalize its psychological operations capabilities and use them to support national security objectives in all legal and proper areas.

Secretary of Defense Caspar W. Weinberger ordered an extensive study on how best to implement the president's direction. This action resulted in a master plan to revitalize military capabilities and better apply them to support long-term peacetime national security objectives. Key elements in his plan involved (1) separating the profession of PSYOP from the more narrow field of special operations and (2) significantly restructuring staffs and units. He also recommended enhanced interagency coordination.

It is important to stress the fact that US military psychological operations in peacetime do not involve misinformation or anything related to the Soviet concept of disinformation. Truth must be our guideline in every undertaking. After World War II, Dick Crossman, the British master of international influence activities, counseled that truth is the very essence of strategic peacetime foreign policy efforts and that it should be our fundamental guideline. US policy must endorse his astute recommendation.

DOD is still engaged in a period of transition to revitalize its military PSYOP capabilities. We envision a number of strategic roles and missions, all of which support objectives derived from published national security policy.

According to the annual defense guidance document, our fundamental national security objective is to deter conflict. This is clearly a psychological phenomenon because it occurs in the mind of a potential enemy. When an enemy perceives that it would be too costly for him to attack or that he would probably lose if he started a war, then he elects not to attack and conflict has been deterred. His mental decision, based upon his perception of our capabilities and resolve, is the key element in the process of deterrence. Military PSYOP in support of this fundamental defense objective, then, should seek to clarify and focus this perception.

Military PSYOP programs designed to enhance deterrence should include those that support arms control talks as well as those that clarify and explain national aims for strategic nuclear force posture. Efforts to clearly describe our policies for peaceful military uses of space and the strategic defense initiative (SDI) can readily be listed as contributing to deterrence. The fact that the term *star wars* was publicly attached to the SDI program aided the Soviets considerably in their propaganda campaigns against it. There were two reasons for this: (1) the term evoked subconscious perceptions of science fiction and fantasy among various public audiences and (2) "wars" is perceived as aggressive rather than defensive. As a result, there was little support for the program. Nevertheless, little was done to decouple this negative terminology. In the future, more must be done to properly support national policy.

A simple thing like terminology can indeed be a major factor in various public audiences' acceptance of—or support for—any policy. Recent history shows that the term *neutron bomb* was a key element in the Soviets' eminently successful campaign a decade ago to prevent President Jimmy Carter from deploying enhanced radiation weapons in Europe. In like manner, we can recall how our national leaders were duped into referring to Iranian terrorists and kidnappers as students at the outset of the Iranian Embassy hostage debacle in 1979. These are clear examples of how terminology can sway public perceptions on a global basis.

Certainly, strategic military PSYOP programs should play a significant role in countering terrorism. In addition to gaining some control over media terminology, we need information programs that will disassociate terrorist groups from pockets of popular support. Such programs should publicly impugn the worth of the terrorists' objectives and denigrate their leadership. We should also seek to reduce their internal cohesion and cause them to waste their resources defensively; for example, in hiding and vetting new members.

Above all, we should seek to positively influence foreign policy efforts to reduce or eliminate foreign governmental support to terrorist groups. With firm evidence that the governments of Libya, North Korea, and Cuba support various terrorist groups, we can and should make publicly supported diplomatic efforts against such governments. Although we have the capability to do this, there are tremendous administrative and bureaucratic hurdles to surmount in developing such programs.

There is a major role to play in countering global disinformation efforts as used by the former Soviet Union. A thoroughly documented Soviet disinformation campaign attributed the cause of acquired immune deficiency syndrome (AIDS) to the Central Intelligence Agency (CIA) and DOD experiments to develop biological warfare weapons. While this was a total falsehood and seemed preposterous to most Americans, the disinformation was found credible by several audiences elsewhere in the world. We are only at the beginning of the level of fear that this epidemic can cause in our world, but the Soviets and others recognized early on that fear inhibits rational thinking and vastly enhances the effectiveness of their propaganda and disinformation.

The second phase of their AIDS disinformation campaign began surfacing in foreign public news items that accused US military personnel stationed overseas of spreading AIDS in various host nations. News stories and leaflets with this theme appeared in the Philippines, Africa, and the Middle East. The obvious objective was to undermine US military presence and influence on a global basis. Still a foreign policy problem, it could be helped by a comprehensive and coordinated foreign policy that vigorously exposes and describes disinformation campaigns while pointing out extensive US efforts to control the disease.

Many are probably not aware that the Soviet Union engaged in a sinister and very dangerous game of economic warfare against the United States. The area of strategic metals provides a good example. Africa was the most visible battlefield in this subtle economic struggle, having as it does an abundance of these resources easily and economically available. The US is about 10 percent self-sufficient (within our borders) in supplies of this type; the USSR was about 90 percent self-sufficient. Metals such as chromium, cobalt, zirconium, titanium, and tantalum are essential to any modern high-technology industrial society. The military dimension is obvious from the extent to which defense systems depend on technology.

The Soviets attempted to keep their supplies in the bank—mostly in Siberia, under the ice. They were content to compete with the US on the world market, thus driving up the price. In various countries of Africa, we saw propaganda, disinformation, and guerrilla warfare drive the price up and the Americans out. In Zaire, this combination of activities caused a 600 percent price increase for cobalt in a two-year time period (1979–81). Now, years later, we are still wondering how to gain control of this issue.

Other areas on the economic warfare front were international financing (manipulating debtor nations and credit extensions), protectionism (still an issue), and peaceful uses of nuclear energy. Propaganda in these areas is increasingly more subtle and less noticed, but it should be monitored. Resulting evaluations should be placed in our national security and foreign policy development processes. And we must ensure that the military dimension is properly coordinated.

There is a military dimension in negotiations on the law of the sea—treaties involving fishing rights, oil drilling, navigation rights, and seabed mining. We also have legitimate national security interests in extradition treaties for terrorists and drug traffickers.

Persuasive communications in peacetime may be accurately likened to advertising and public relations programs. Secrets to success involve getting the facts out in clear and understandable ways, relating these facts to perceived human needs, and ensuring enough repetition so that the message is received and reinforced. Various target audiences have differing perceived needs, scripts, and priorities; therefore, the same message must be packaged in different ways in order to appeal to a wider spectrum.

The United States government has a wide variety of communication channels. Each major government agency has offices whose purpose is to announce its policies and explain how its activities support national policies and goals. The United States Information Agency is specifically charged with informing peoples in foreign nations about US policies and describing American culture and our democratic way of life.

This is done through speakers and seminars, press conferences, and various publications. Radio and television programming, such as the "Voice of America" and the new "Worldnet," are able to reach vast audiences.

Areas where there is the most room for improvement are in (1) coordinating various government agencies and (2) packaging the message to fit the audience. Often in crisis situations or periods of tension, the need for rapid response inhibits coordination among various branches of the government. Officials tend to couch their statements in terms they are comfortable with and to speak from their personal points of view, neglecting to consider the cultural aspects and aspirations of foreign audiences. Improvement in these areas is absolutely essential for more effective communications.

Repetition is also necessary. Government officials are busy people; they tend to make policy statements once, and then get on to the next item of business. But policy statements don't necessarily have a life of their own. Unless repeated in various ways and related anew to emerging situations, they tend to submerge in piles of records and gather dust in archives.

The Soviets clearly demonstrated that the most blatant lies can be successfully put across through massive repetition of disinformation and propaganda. In fact, repetition was the secret of their success. Some Soviet propaganda lines were constantly repeated for 50 years; although there was not a single economic success story arising out of a communist or socialist system, belief in this big lie continued over much of the world. People are still willing to give up their freedom for a dream that is inconsistent with the human condition. This is the threat we face from the competition.

Initial changes by Mikhail S. Gorbachev to the Soviet global propaganda machine reflected a stepped-up effort to supplant US power and influence. After the Twenty-seventh Party Congress in the spring of 1986, their propaganda apparatus was restructured and even more centralized. The International Information Department was disestablished and its responsibilities were consolidated within the International Department (ID) and the Propaganda Department (PD). The PD had been responsible for internal domestic propaganda. After the changes, it had oversight for several activities oriented toward foreign audiences; for example, Tass and Novosti as well as Radio Moscow. This move reflected the theory that foreign propaganda and domestic propaganda are intimately linked, and that greater efficiency can be gained through centralization.

At the same time, Gorbachev's reforms (under his program of glasnost) were being widely touted in the West. Many were led to believe that this change represented a welcome softening of formerly hard Soviet propaganda lines. In fact, one of the major propaganda efforts to sugarcoat this program sought to convince people that the term *glasnost* meant "openness." A more accurate translation would be "publicity." Proper translation is not simply a semantic game; it shaped the debate on effective policy response toward the USSR.

Most analysts feel that the real intent of glasnost was not to create an open society, but to consolidate Gorbachev's power by modernizing the system and ending economic stagnation. Carefully staged events, such as the Moscow Peace Forum, dissident releases, rock concerts, and television shows, sought to take advantage of easily co-opted Western participants in order to improve external opinion of the Soviet Union, further Soviet propaganda lines, and secure badly needed Western technology.

Those events should have been viewed with extreme distrust, inasmuch as they represented a quantum leap in Soviet propaganda sophistication. We saw significant improvements beginning under Yuri V. Andropov's administration, with expanded use of professional psychology and psychiatry along with infusions of equipment and talent that increased the quality of Soviet PSYOP. Under glasnost, we watched a merger between these professional propagandists and the innocent Western media. It was and is a dangerous and threatening evolution.

Still, the Soviets lost control of glasnost, affording an opportunity to project democratic messages to various Soviet audiences while those channels were open.

If US government agencies can improve their coordination and persuasive communications capabilities, we may be able to slow and then reverse our losing trend in exerting world influence. Propaganda and disinformation that seek to cause people to trade freedom for lies is a most insidious and dangerous threat, and freedom is far too precious a thing to be defended by haphazard efforts. We do have one significant advantage: There is no need to lie because properly packaged truth is the very best propaganda.

No More Tactical Information Detachments

US Military Psychological Operations in Transition

Col Alfred H. Paddock, Jr., USA, Retired

US military PSYOP today faces an uncertain future. This paper provides a brief perspective on the evolution of US military PSYOP during the twentieth century, examines the conditions that led to a revitalization of these capabilities during the 1980s, contrasts this revitalization with the international and domestic environment faced by the Department of Defense (DOD) today, and outlines some of the major challenges for the US military PSYOP community during a transitional period.[1]

A broad definition of psychological operations details the planned use of communications to influence human attitudes and behavior. They consist of political, military, and ideological actions conducted on target groups to create desired behavior, emotions, and attitudes. If used properly, PSYOP will precede, accompany, and follow all applications of force. The military should carry out this course of action under the broader umbrella of national policy. In addition, it should coordinate fully and carefully the military component of the overall PSYOP effort with other agencies of government.

More specifically, PSYOP can demoralize, disorient, and confuse hostile groups. Against such groups, psychological operations are employed as offensive weapons to enhance the overall effectiveness of military operations. They also can unite, inform, and bolster the morale of nonhostile groups. When targeting neutral or friendly groups, they are used to support military objectives by developing cooperative attitudes and behavior in the targeted group.

Evolution of US Military
Psychological Operations

The level of interest in military psychological operations during this century has been episodic, basically rising and falling during and after the major conflicts in which US forces have been committed. Over this period, most of the activity in military PSYOP centered on the Army. The following brief historical perspective, therefore, focuses on the Army's experience to illustrate the fortunes of military PSYOP.

While giving psychological warfare (psywar) only token recognition in World War I, the Army established both the Psychological Warfare Subsection of G-2 in the War Department and the Propaganda Section of G-2 in General Headquarters, American Expeditionary Forces. Military tactical psychological warfare centered on the production of leaflets; radios did not exist as a means of communication, and loudspeakers were primitive. Military propaganda concentrated on producing surrender appeals; balloons and airplanes were the primary method for their dissemination.

From 1918 to 1941, no psychological warfare office existed in the War Department. The lessons of experience were lost, and by 1941 only one officer on the War Department staff had had psychological warfare service in the previous war.

During World War II, most of the Army's operational work in psywar took place at the theater level, where the responsible organization was normally designated a psychological warfare branch (PWB). The largest of these organizations at the theater level, the PWB at Allied Forces Headquarters (PWB/AFHQ), was activated in North Africa in November 1942 at the order of Gen Dwight D. Eisenhower. It was expanded in February 1944 to the Psychological Warfare Division, Supreme Headquarters, Allied Expeditionary Force (PWD/SHAEF). Psychological warfare was defined as "the dissemination of propaganda designed to undermine the enemy's will to resist, demoralize his forces and sustain the morale of our supporters."[2]

The basic Army field operating unit for psywar was the Mobile Radio Broadcasting (MRB) Company. The equipment for this company was unlike anything conventional soldiers had seen in the field: public address systems, radios, monitoring sets, loudspeakers, typewriters, mobile printing presses, and leaflet bombs. MRB units were usually divided by the separate Army groups and field armies into small teams, often to work in direct support of frontline conventional combat units. Five such units eventually served under PWD/SHAEF. Although these units resulted from improvisation in 1943 and 1944, the doctrinal and organizational concepts they embodied reappeared in the psychological warfare units formed during the Korean conflict.

During 1945–46, Army psychological warfare staffs and units dissipated with the general demobilization of the military establishment. Despite the efforts of a few senior civilian and military officials to retain a military PSYOP capability, the Tactical Information Detachment was the only operational psychological warfare troop unit in the US Army when the North Koreans attacked South Korea in June 1950. Organized at Fort Riley, Kansas, in 1947, this detachment had been reorganized as the 1st Loudspeaker and Leaflet (L&L) Company. Sent to Korea in the fall of 1950, it served as the Eighth Army's tactical propaganda unit throughout the conflict. Tactical propaganda, sometimes called combat propaganda, was directed at a specific audience in the forward battle areas and used in support of localized operations. Mobile loudspeakers mounted on vehicles and aircraft became a primary means of conducting tactical propaganda in Korea.

To conduct full-scale strategic operations, the 1st Radio Broadcasting and Leaflet (RB&L) Group was organized at Fort Riley and shipped to Korea in July 1951. The 1st RB&L Group was specifically designed to conduct strategic propaganda in direct support of military operations. This propaganda was intended to further long-term strategic aims and was directed at enemy forces, populations, and enemy-occupied areas. The 1st RB&L Group had the equipment and capability to produce newspapers and leaflets, and to augment or replace other means of broadcasting radio propaganda. The group supervised a radio station network, known as the "Voice of the United Nations," and often produced more than 200 million leaflets a week that were disseminated by aircraft or by specially designed artillery shells. The leaflets expressed various themes. Some leaflets, for example, offered inducements for enemy soldiers to surrender; others bolstered the morale of Korean civilians by proclaiming UN support.

Although the RB&L Group was a concept accelerated to meet the requirements of the Korean conflict, it performed functions similar to those used in psychological warfare in World War II. Its MRB Company bore a direct linkage to the mobile radio broadcasting companies formed under PWD/SHAEF to conduct operations in North Africa and the European theater during 1944–45. However, the MRB companies were organized during World War II to perform tactical psywar; radio later became an essentially strategic weapon that had no place in a purely tactical psychological unit. The strategic concept embodied in the RB&L Group was destined to figure prominently in the psywar capability that was subsequently formed as part of the Psychological Warfare Center in 1952. The tactical idea expressed by the L&L Company also influenced the capability that was developed there. As originally established at Fort Bragg, North Carolina, the Psychological Warfare Center consisted of a psychological warfare school, the 6th Radio and Broadcasting Group, a psychological warfare board, and the 10th Special Forces Group. The mission of this unprecedented center:

> To conduct individual training and supervise unit training in Psychological Warfare and Special Forces Operations; to develop and test Psychological Warfare and Special Forces doctrine, procedures, tactics, and techniques; to test and evaluate equipment employed in Psychological Warfare and Special Forces Operations.[3]

After an initial burst of activity fueled by the Korean conflict and by fears of a possible outbreak of war in Europe, interest in the Psychological Warfare Center began to dissipate. Its title was changed to the Special Warfare Center in 1956, but the Army's psychological operations capability had eroded by the early 1960s.

Consequently, an insufficient base of PSYOP-trained officers were available when the 6th Psychological Operations Battalion was activated in Vietnam in 1965. By 1967, the Army's PSYOP forces in Vietnam had been expanded to a group (the 4th) with four battalions, one in each of the four corps tactical zones (CTZ). The group served under the commander, US Military Assistance Command, Vietnam, with the J-3's Psychological Operations Division exercising direct staff supervision. The Joint US Public Affairs Office (JUSPAO) provided US PSYOP policy guidance not only to the civilian agencies but also to all military PSYOP elements.

In addition to providing tactical support to field force commanders, the 4th Psychological Operations Group assisted the South Vietnamese government in its communication effort down to the hamlet level. The group headquarters operated a 50,000-watt radio station and high-speed heavy printing presses, published a magazine for Vietnamese employees working for US government and civilian agencies, and could research and develop propaganda materials.

In contrast, PSYOP battalions had light printing presses, a research and propaganda development capability, personnel to work with the US Air Force Special Operations units for aerial leaflet and loudspeaker missions, and ground loudspeaker and audiovisual teams. Loudspeaker and audiovisual teams operated with US divisions and brigades or with province advisory teams. The 7th Psychological Operations Group in Okinawa provided valuable backup support in printing and in disseminating leaflets at high altitudes.

During the height of US involvement in Southeast Asia, the Army stationed PSYOP units at Fort Bragg and in Germany, Panama, and Okinawa. These units were established in addition to the 4th Psychological Operations Group in the Republic of Vietnam. By the mid-1970s, however, all that remained in the active component was an inadequately staffed and antiquatedly equipped group at Fort Bragg—a condition that did not improve significantly for 10 years.

Revitalization of PSYOP in the 1980s

The 1980s saw an upturn in the fortunes of military PSYOP, made possible by several conditions. President Ronald Reagan, who assumed office with a strong anticommunist orientation, characterized the Soviet Union as the "evil empire" early in his tenure. He directed that a program be developed to support noncommunist insurgencies (the "Reagan Doctrine") around the world. With public opinion supporting him, President Reagan embarked upon a large buildup of the US defense establishment—a buildup that benefited military PSYOP.

The president then outlined, in a number of national security directives, his approach to the psychological dimension of national power. His initial national security strategy, announced in the summer of 1981, contained four basic components: diplomatic, economic, military, and informational. This emphasis on the psychological component was retained in subsequent Reagan national security strategies.

In January 1983, President Reagan signed National Security Decision 77, "Management of Public Diplomacy Relative to National Security." This directive defined public diplomacy rather broadly, stating that it "is comprised of those actions of the U.S. Government designed to generate support for our national security objectives." The term generally evolved to embrace a broad range of informational and cultural activities. The decision also established an interagency mechanism to plan and coordinate public affairs, information, political, and broadcasting activities of the US government.

In early 1984, the president directed DOD to rebuild its military PSYOP capabilities. In response to this directive, Secretary of Defense Caspar Weinberger launched a major evaluation of the department's capabilities and needs in psychological operations. This evaluation concluded that DOD's PSYOP capabilities had been allowed to atrophy over the previous decade. Across-the-board deficiencies had developed in policy guidance, roles and missions, doctrine, organization, force structure, operational concepts, planning, programming, training, logistics, intelligence support, readiness, personnel programs, and—most importantly—attitude, underscoring the need for education and heightened awareness at all levels of military and civilian organizations.

DOD PSYOP Master Plan of 1985

Secretary Weinberger elected to use a DOD PSYOP master plan as the framework for rebuilding military PSYOP capabilities. Approved in mid-1985, the plan served as a comprehensive design for the fundamental improvement of the department's capabilities to effectively perform worldwide psychological operations in support of national objectives in peace and crisis and at all levels of conflict.

The plan specified a number of remedial actions—over 200, in fact—to be implemented over several years. Several essential themes manifested themselves.

The first theme highlighted the need to develop a comprehensive joint doctrine for the formulation, direction, coordination, and conduct of PSYOP in peace, crisis, and war. In effect, this doctrine provided a foundation for the revitalization effort. Among other things, it sought to enunciate the function of PSYOP as a force multiplier in all military activity, establish the conceptual framework for planning and implementation, and delineate roles and responsibilities of the several components. The Joint Chiefs of Staff (JCS) published this doctrine in 1987. It was later revised to reflect US experience with PSYOP in support of military operations in Panama and the Persian Gulf.

The development of doctrine was to be paralleled by major improvements in PSYOP planning, the second theme of the 1985 Master Plan. It devoted insufficient human talent to full-time, meaningful, sustained PSYOP planning at appropriate staff levels. The report card on this major deficiency offers mixed reviews, but some evidence of progress exists. Creation of a small psychological operations directorate, the first such office to exist in the Office of the Secretary of Defense (OSD) in over 20 years, indicated the seriousness with which Secretary Weinberger undertook the revitalization effort. The PSYOP staff element in the Office of the Joint Chiefs of Staff (OJCS) was upgraded from a branch to a division. It also showed an increase in personnel, but later reverted to a branch and lost a few personnel spaces. The Department of the Army staff created a psychological operations and civil affairs division, where before only one full-time officer had been assigned to this activity.

Outside the Pentagon, a few changes occurred. Among the unified commands, United States Southern Command (USSOUTHCOM) created a PSYOP detachment to augment its staff capability. The US Special Operations Command (USSOCOM) was established in 1987 at MacDill Air Force Base, Florida. Included in this unified command staff was a directorate for psychological operations and civil affairs (J-9). More recently, in 1990, the Army created the Civil Affairs and Psychological Operations Command at Fort Bragg. Commanded by a USAR brigadier general and staffed primarily by reserve component personnel on varying periods of active duty, this headquarters of 146 personnel (later reduced to 80) serves all active and reserve component Army civil affairs and psychological operations units. This new organization reports to the US Army Special Operations Command (USASOC) at Fort Bragg.

Much more remains to be done in this area, however. Perhaps the key requirement is that staff officers receive formal training in this specialized area and that psychological operations become an integral part of the operational course of action in any plan. A few innovative and resourceful staff officers can make a vast difference if they have access to a command's major planning activities. Because of the paucity of trained PSYOP staff officers in major commands, the burden for planning too often falls upon the Army's 4th Psychological Operations Group, the only active duty unit of its type among the services.

Development of adequate numbers of PSYOP planners is where the Air Force, Navy, and Marine Corps can make an important contribution to the revitalization envisaged in the 1985 Master Plan. The conduct of psychological operations is not the exclusive domain of such specialized units as those in the Army. In the current resource-constrained environment, only a small possibility exists that the other services will opt to field new PSYOP units. On the other hand, there is every reason to expect those services to develop fully qualified PSYOP planners to meet their own needs and to provide their proportionate share on joint staffs.

To help address this need and in response to the 1985 Master Plan, the Army's John F. Kennedy Special Warfare Center and School at Fort Bragg developed a joint PSYOP staff planning course in 1988. A large proportion of its students have been from the Army's reserve component PSYOP community. Too few representatives from the other services have participated, and it is uncertain how many of them actually were assigned to staff positions where their expertise could be utilized. Reportedly, the Army plans to eliminate the course in 1996. If it does, this will be unfortunate—it is the only formal joint PSYOP planning course currently available. What is needed instead is a renewed and continuous emphasis that would enable the course to fulfill its original purpose. This will not happen, however, until the services begin to send more officers to the course and ensure their assignment to positions where their training can be utilized.

Closely related was the need to educate our officer corps on psychological operations, the third theme. As indicated earlier, the 1985 Master Plan stated that the root cause of the atrophy of our military PSYOP capabilities was a lack of understanding of psychological operations, their value, and their application. Some improvement was made in this critical area as a result of frequent briefings to senior commanders and staff officers by PSYOP personnel, the professionalism of PSYOP units in contingency planning and in support of conventional forces on joint training exercises, and the steady improvement of PSYOP studies and assessments in support of the unified commands and national level agencies. Certainly, the credible performance of PSYOP forces during combat operations in Panama and the Persian Gulf also played a key role in this regard.

Additionally, the PSYOP courses being presented by both the Air Force and the Army help to address the deficiencies in this area of PSYOP awareness and understanding. The Air Force's Joint Senior Psychological Operations Course, held four times a year by their Special Operations School at Hurlburt Field, Florida, provides selected senior officers and civilians with an awareness of how psychological operations can support national objectives throughout the spectrum of conflict. This course initially showed considerable promise as an educational tool, but it attracts too many personnel whose duties have little relationship with psychological operations. Again, this situation requires renewed and continuous emphasis to interested senior officials in key positions. Unfortunately, these personnel too often find it difficult to fit a three-or four-day course away from the office into their busy schedules. A shorter orientation—perhaps four hours, presented by a traveling team—should be considered as an addition to the present course.

As was the case before the Vietnam War, the 1985 Master Plan stated that PSYOP instruction in our mainstream service school system—where our future commanders and staff officers are trained—was limited or nonexistent. This situation not only made more difficult the PSYOP community's job of educating supported units on this unique weapons system—it also had a negative effect when priorities concerning force modernization were being set.

To address this loss of PSYOP institutional memory, the under secretary of defense for policy, Dr Fred Ikle, sent letters in 1986 and 1987 emphasizing the secretary of defense's intention to revitalize our military PSYOP capabilities and offering an OSD presentation on psychological operations to commandants of the colleges and the command and staff colleges. The intent was to stimulate interest in a more extensive treatment of PSYOP in service school curricula. The military services were asked to review all levels of their professional military education (PME) curricula, both officer and enlisted, and to develop new educational goals that would upgrade their psychological operations PME. As a second part of this approach, OSD initiated a contract to develop applicable PSYOP curriculum materials for the services.

These materials were in fact provided to the services, but there is little evidence to suggest that the services integrated PSYOP into the curricula of their schools. Nor have great speakers made presentations on PSYOP to service schools in a coherent and consistent manner (some presentations are made in electives). Institutionalization of PSYOP understanding will not occur until service schools have removed this deficiency.

The fourth broad theme in the 1985 Master Plan encompassed the need to modernize our PSYOP force structure in terms of both personnel and equipment. With the exception of an increase of the 4th Psychological Operations Group's strength, the force structure today is essentially the same as in 1985. The Navy has an excellent radio and television capability in its reserves and a 10-kilowatt (kW) mobile radio transmitter that is assigned to its Tactical Deception Group (Atlantic), both of which can be used to support PSYOP activities. The Air Force has a National Guard squadron of specially fitted C-130 aircraft for radio and television support of PSYOP along with other duties. Also, a handful of Air Force officers with PSYOP expertise are serving in key positions in the Pentagon, among the unified commands, and at their Special Operations School at Hurlburt Field. The Marine Corps has two civil affairs groups with PSYOP as a secondary mission in its reserves. Only the Army has active duty forces dedicated solely to psychological operations.

The 4th Psychological Operations Group at Fort Bragg is all that remains of the Army's active PSYOP capability. Assigned to the US Army Civil Affairs and Psychological Operations Command, and to a part of the US Army Special Operations Command, the 4th Group has many worldwide missions and responsibilities. The 4th provides support to all levels of DOD, from unified command through division. It supports both conventional forces and special operations forces. In addition, it is often called upon to provide support to national level agencies and organizations. If any one military unit can be adjudged a "national asset," surely the 4th Psychological Operations Group fits the requirements.

Essentially, a military PSYOP unit engages in two broad activities: (1) research/analysis and (2) operations. The first activity consists of continually monitoring and assessing psychological environments in specific foreign nations to determine how those situations affect the formulation and execution of US policies and actions. This research and analysis results in published studies and assessments that are unique; they provide the foundation for establishing psychological objectives that support US goals related to foreign nations or groups. Research and analysis is therefore essential to accomplishment of the second broad activity: planning and executing PSYOP campaigns that employ communications media and other techniques designed to cause selected foreign groups and individuals to support US national and military objectives.

In peacetime, a military PSYOP unit conducts research and analysis of specific geographic regions and target audiences, develops PSYOP plans to support conventional and special operations units, and participates in field exercises that employ these plans. Because of the paucity of PSYOP expertise at unified commands, the 4th Psychological Operations Group provides staff assistance and advice to headquarters and to other major commands.

It should be eminently clear from the foregoing that one active duty PSYOP organization cannot support all unified command requirements for mid- or high-level intensity conflicts. The reserves are, therefore, a vital component of the "PSYOP community," since 73 percent of the Army's PSYOP mobilization capability lies in its reserve component units. Serving as the Army's planning agent to align the AC and RC units program (CAPSTONE, which links reserve units with the units they would support upon mobilization), the Civil Affairs and Psychological Operations Command (USACAPOC) provides training assistance to, and coordinates the contingency planning efforts of, reserve units. The 4th Group assists USACAPOC in this effort.

Generally speaking, then, USACAPOC and the 4th Psychological Operations Group act as a "strategic nucleus" for the PSYOP community; they provide the bulk of peacetime and low-intensity conflict requirements, give direction and guidance to the PSYOP community for contingency planning and participation in peacetime exercises, and serve as the command and control nucleus for generally or partially mobilizing reserve component forces. The reserve component performs its planning and training responsibilities under the CAPSTONE program and prepares for general or partial mobilization in support of the unified commands.

Paradoxically, the successful CAPSTONE program underscores one of the PSYOP community's most glaring weaknesses: its limited capability to respond to peacetime and low-intensity conflict requirements. While mid- and high-intensity conflict requires either partial or general mobilization of the reserve component, the active component must provide most PSYOP activities during peacetime and at the lower end of the conflict spectrum.

As a result of initiatives undertaken in the early 1980s, the Army increased the strength of the 4th Group by 500 personnel, roughly doubling its size by the mid- to late 1980s. The 1985 DOD PSYOP Master Plan provided additional impetus for this increase. The reserve component PSYOP structure has remained relatively stable in recent years, with a little over 3,000 personnel.

In terms of personnel quality, the Army's military occupational specialty (MOS) for PSYOP enlisted personnel was a welcome accomplishment. Reports of the high quality of personnel being trained under this specialty have been encouraging.

Less encouraging was the Army's decision in the mid-1980s to remove its officer PSYOP and civil affairs MOS from the foreign area officer (FAO) specialty. The change was disturbing because it separated psychological operations from the specialty that had provided its intellectual lifeblood. The core of the area expertise (knowledge of foreign cultures) and the analytic capability of psychological operations fell within the FAO specialty.

The Army's initial decision included both PSYOP and civil affairs in the special operations functional area. Subsequently, however, the Army developed separate functional areas for PSYOP (FA 39B) and civil affairs (FA 39C). It remains to be seen whether this decision will provide the quality officers that previously had been produced by the FAO program.

With respect to the modernization of psychological operations equipment, the progress has been more positive. Several Army initiatives in the early and mid-1980s upgraded active and reserve component print, radio, loudspeaker, and audiovisual capabilities. The 4th Group's media production center, built in the mid-1980s, represented a quantum leap in the military's PSYOP capability to operate in a modern audiovisual environment. Similarly, the Air Force allocated funds to modernize its National Guard aircraft dedicated to support PSYOP.

Nevertheless, modernization of PSYOP-unique equipment requires continued emphasis. For example, with one exception, the old, relatively immobile 50,000 kW radios (the TRT-22s) in the PSYOP community have not been replaced. PSYOP loudspeakers once again need upgrading, and the Air Force needs a device that can rapidly release leaflets in high-altitude operations.

Two of the most controversial themes of the 1985 DOD PSYOP Master Plan were the organizational separation of psychological operations from special operations and its corollary, creation of a joint psychological operations center. Indeed, these were the only truly revolutionary initiatives in the original plan—and both failed.

The authors of the 1985 Master Plan believed that, in general, the subordination of psychological operations to special operations detracted from recognition of the overall applicability of PSYOP in times of peace, crisis, and war. The

1985 Master Plan, therefore, called for separating psychological operations from special operations throughout DOD, including departmental levels and all headquarters and staffs of the JCS, the services, the unified and specified commands, and the component and subordinate commands. The plan outlined several reasons for separating PSYOP from special operations.

Planning, particularly in the unified and specified commands, suffered because the sole PSYOP planner was usually located in the special operations staff element and was employed only part-time in psychological operations. This subordination detracted from the broader responsibility of planning psychological operations support for the theater's total requirements, particularly in those missions that link military psychological operations and national objectives, policy, and strategy, and that require in-theater interagency cooperation.

Further, the subordination of psychological operations units to the special operations field, including its command and control structure, contributed to the lack of understanding of PSYOP, its uses and capabilities, by military officers and senior civilians. This association and subordination caused many in the field to conclude that psychological operations were focused primarily in support of special operations missions.

While PSYOP does have a mission in support of special operations, it also has a much broader application in peacetime and in crisis—with or without accompanying military operations—and across the entire spectrum of conflict. In the 1980s, only 10 percent of the Army's psychological operations forces—active and reserve—were designated by contingency plans to support special operations units in wartime.

Similarly, the argument for separation of PSYOP from special operations was implicit in the Army's and JCS's arrangements for wartime command and control of psychological operations. In wartime, most of the PSYOP forces are aligned (within the unified commands) with a chain of command that is totally separate from that of special operations forces. PSYOP units are combat support forces and are employed at both strategic and tactical levels from the theater to the division as a matter of routine; special operations forces are employed primarily as strategic assets only on an exceptional basis.

Because of its controversial nature, the separation directed by the 1985 Master Plan was slow in implementation. To be sure, some important initial steps were taken. A new OSD psychological operations directorate established in January 1986 was assigned to the principal deputy under secretary of defense for policy and was separated from special operations in terms of overall policy responsibilities. On the joint staff, PSYOP was removed from the Joint Special Operations Agency and placed under the J-3. An important precedent was established on the Department of the Army staff when a separate PSYOP and civil affairs division was created; and two unified commands—USSOUTHCOM and USCENTCOM—effected the separation (PSYOP from special operations) within their staffs.

But for the large part, separation did not take place among the unified commands, subordinate service headquarters, or at the operational level. The Army, for example, took the position that separation applied only to staff levels, not forces. A major reason for the delay, of course, was the uncertainty resulting from the congressionally mandated reorganization of special operations forces. This subject will be discussed later in more detail. Suffice it to say at this point that full separation did not occur.

The separation issue was closely related to the final major theme of the original master plan: creation of a joint psychological operations center. According to the master plan, psychological operations was sufficiently important to warrant the creation of a separate center dedicated to the long-term development and nurturing of this unique capability. This center eventually became the organizational and intellectual font of PSYOP within the Department of Defense.

The responsibilities for the center envisaged in the 1985 Master Plan included long-range strategic psychological operations plans; continuing education and training of personnel research and analytical studies; and development of equipment. The center was also to assist OJCS and OSD to develop, plan, and coordinate the defense portion of interagency activities, and to assist the unified and specified commands in their planning of psychological operations. To best accomplish these tasks, the plan favored the Washington, D.C. area for the center's location.

The joint psychological operations center was to consist of two separate but mutually supporting elements, one operational and the other developmental, each indispensable to the other. These elements were to have been the 4th Psychological Operations Group along with PSYOP spaces in the Army's John F. Kennedy Special Warfare Center and School. This concept would have used existing PSYOP units and personnel spaces to provide the nucleus of initial manpower requirements.

The plan indicated that initially the center probably would be under Army management because the other services had few personnel to contribute. However, as representation from the other services increased to more than token level, the joint character of the center was to be emphasized by making its command rotational among the services. As with the separation theme, however, the joint psychological operations center envisaged by the original plan met with fierce resistance, particularly from the Army—which, of course, would have been the most heavily affected by the initiative.

Implementation of both initiatives—the separation of PSYOP from special operations and the creation of a joint psychological operations center—became, in effect, hostage to the resolution of several issues that resulted from the congressionally mandated reorganization of special operations forces. In October 1986, Congress passed a special operations force (SOF) reform package sponsored by Senators William Cohen (R.-Maine) and Sam Nunn (D.-Ga.). The law created a unified SOF command under a four-star general or flag officer, an assistant secretary of defense for special operations and low-intensity conflict, and a board for low-intensity conflict within the National Security Council (NSC). It recommended the appointment of a deputy assistant to the president for low-intensity conflict. Particularly important, it directed the secretary of defense to create a major force program category for DOD's five-year defense plan for special operations forces.

After passage of this legislation, a major question arose on whether to include PSYOP and civil affairs units in the assignment of forces to the new USSOCOM. Good arguments were presented on both sides of this issue. Two of the major advantages advanced for inclusion were that these forces would benefit from the sponsorship of a unified commander and participation in the major force program for SOF. But the secretary of defense's previous guidance to separate PSYOP from special operations—as part of his decision concerning implementation of the 1985 DOD Psychological Operations Master Plan—had to be considered. After a lengthy review of this issue, Secretary Weinberger decided to assign Army and Air Force active and reserve component psychological operations and civil affairs units to USSOCOM, effective 15 October 1987.

Secretary Weinberger's decision, of course, sealed the fate of the two most controversial and far-reaching initiatives of the original plan: separation of PSYOP from special operations and creation of a joint PSYOP center. Separation at the operational level had never taken place, since the Army's PSYOP forces remained under their special operations command at Fort Bragg. And separation at the staff level—which had seen some initial progress, at least in the Pentagon—would inevitably fail as well. On the joint staff, PSYOP was transferred back to special operations. And in May 1992, overall policy responsibility for PSYOP in OSD was moved to the Office of the Assistant Secretary of Defense (OASD) for Special Operations and Low-Intensity Conflict.

The secretary's decision also made moot the creation of a joint PSYOP center, as originally conceived in the 1985 Master Plan. The forces and personnel spaces scheduled for the nucleus of this project remained under the Army's special operations command and as part of USSOCOM. Even the creation of a much smaller version of the joint PSYOP center—located in the Washington, D.C. area and concentrating primarily on strategic PSYOP—has become unlikely in the current environment.

A candid assessment of the 1985 plan, therefore, provides mixed reviews. The most significant initiatives on increasing the strength of the 4th Psychological Operations Group—establishing an enlisted PSYOP MOS, modernizing PSYOP-unique equipment for both active and reserve component units, and undertaking detailed PSYOP contingency planning in support of the unified commands—were all undertaken by the Army in the early 1980s, prior to the 1985 plan. To be sure, the plan provided additional emphasis to help bring these initiatives to fruition, but the initial impetus came from the Army. Improvements undoubtedly occurred in the formulation of joint doctrine, indoctrination of the officer corps, and staff planning, but these areas still require attention. And the two truly revolutionary themes of the original master plan—separation of PSYOP from special operations and creation of a joint PSYOP center—did not become a reality. In the final analysis, perhaps the greatest benefit to the PSYOP community was the morale-enhancing evidence of senior-level interest, due in large part to the dedication of the late Gen Richard G. Stilwell, USA, Retired, who spearheaded the 1985 plan while serving as the deputy under secretary of defense for policy.

Nevertheless, while the plan's ambitious goals were not fully realized, the revitalization efforts of the 1980s made possible the highly credible performance of PSYOP forces in Panama (Operations *Just Cause* and *Promote Liberty*) and the Persian Gulf (Operations *Desert Shield* and *Desert Storm*). The early integration of PSYOP into military planning for these operations—heretofore all too often a serious deficiency—contributed significantly to its successful utilization. In both contingencies, PSYOP forces employed a wide range of media, including loudspeaker broadcasts, radio transmissions, posters, and leaflets. Principal PSYOP objectives were to (1) cause the enemy to cease resistance and surrender, (2) solicit information and the turn-in of weapons, (3) inform civilians and keep them out of the battle areas, (4) establish a favorable image of friendly forces, (5) deflect hostile propaganda, and (6) support civil affairs units in emergency relief and consolidation activities. Overall, the goal was to enhance combat operations and reduce casualties on both sides. Successful accomplishment of this goal would have been doubtful without the increased personnel strength of the 4th Psychological Operations Group and the modernization of equipment for both active and reserve component PSYOP units brought about by revitalization.[4]

Thus, the revitalization of military PSYOP capabilities that took place during the 1980s was fueled by a presidentially inspired public perception of the common threat (i.e., the Soviet Union), a "defense consensus" which enabled massive increases in military budgets, several key Reagan administration national security directives that provided impetus for DOD, and the "top-down" emphasis of Secretary Weinberger and a few key advisors which resulted in the 1985 DOD PSYOP Master Plan.

The Changing International and Domestic Environment

A comparison of these earlier conditions with the present environment provides some rather stark contrasts. The most dramatic change has occurred in the American public's perception of the threat.

An important stimulus for this change in attitude was the emergence of Mikhail Gorbachev as Soviet leader in 1985. His domestic political and economic reforms, the withdrawal of troops from Afghanistan, the cessation of "Voice of America" jamming, his announced intention to reduce Soviet military forces by 500,000 men and cut their budget by 14.2 percent, and the intermediate nuclear forces arms control agreement were just some of the actions that helped to lessen international tensions. Polling data of US public attitudes depicted a startling change in just over a two-year period; by July 1988, 94 percent of voters believed that US-USSR relations were stable or getting better. Mr Gorbachev's policies led Margaret Thatcher to conclude in November 1988 that the cold war had come to an end.

If onlookers thought Thatcher's conclusion a bit premature, they should consider that the subsequent years have been truly mind-boggling. The dissolution of the Warsaw Pact and the liberation of Eastern Europe, the reunification of Germany and its emergence as a major power, and the disintegration of the former Soviet Union have been revolutionary geopolitical changes that few could have foreseen. Russian Republic President Boris Yeltsin's signing of a historic strategic arms control agreement with President George Bush in June 1992 is only the most recent indicator that the cold war as we knew it has, in fact, ended.

During the same period that these momentous events have occurred in the international arena, Americans have become more preoccupied with domestic concerns. A lingering economic recession and high jobless rate, concerns over the burgeoning federal budget deficit, and deep-rooted problems of

43

the cities (as vividly seen in the Los Angeles riots) have taken their toll. In February 1992, the conference board's measure of consumer confidence registered its lowest level in 17 years; and in April, four-fifths of Americans surveyed said they believed the country was seriously on the wrong track.[5] These domestic anxieties, coupled with the absence of a clearly discernible external threat, have led to growing pressures for steeper cuts in the defense budget and in foreign aid. The "defense consensus" of public and congressional support during the early Reagan administration has disappeared.

Thus, the context within which defense planning takes place is in transition, as William Hyland, editor of *Foreign Affairs*, reminds us:

> This time the cliché is true; this is a new era, but we are only in the opening phases. It is fruitless to search for a politically correct concept of the national interest to justify American foreign policy. Debating in these categories is itself an intellectual hangover from the Cold War. . . . The US defense posture is only in the preliminary stages of the restructuring that will follow the end of the Soviet threat.[6]

Challenges for Future US Military Psychological Operations

The challenges for military psychological operations are formidable. To be sure, inclusion of PSYOP forces in USSOCOM provides some assistance in meeting these challenges: a four-star proponent (the commander, USSOCOM) and, ostensibly, full participation in the major force program (MFP-11) for SOF. The clout of an assistant secretary of defense (ASD/SOLIC) also should benefit psychological operations at the policy level.

Nevertheless, three broad areas must be addressed if the PSYOP community is to emerge from this period of transition to play a viable role in our nation's security: (1) education of senior civilian and military officials on the value of the psychological dimension; (2) planning for the use of military PSYOP in a changing world; and (3) continued modernization of the PSYOP force structure.

Planning should begin now to ensure that the psychological dimension receives appropriate attention by the new administration—and OASD/SOLIC should play a leading role.

Unfortunately, a loss of momentum has occurred in this area. The national security strategy statements of President Bush (March 1990 and August 1991) placed a much less explicit emphasis on the informational element of national power than did those of Ronald Reagan. The Reagan administration directives that created interagency public diplomacy mechanisms, and that played a vital role in revitalizing DOD's psychological operations, have lain dormant. The NSC has paid little attention to this area.

A review of our national security strategy suggests that OSD, OJCS, and USSOCOM must continue to coordinate their efforts to have psychological operations considered by the NSC. DOD must intensify its efforts to educate new players on the NSC staff and to build allies among the key agencies in order to resuscitate existing national security directives—or create new ones—that provide guidance to psychological operations and public diplomacy activities. To address a long-standing deficiency, DOD sorely needs a senior-level NSC mechanism to integrate and direct the information and psychological operations activities of the various agencies. Such a forum also would provide a vehicle to overcome the differences in operational terminology that have plagued past efforts to integrate interagency planning in these areas.

To meet its needs in a changing world, DOD should undertake once again to update its bilateral agreements with the United States Information Agency (USIA), the Board for International Broadcasting, and the Central Intelligence Agency (CIA). With diminishing budgets almost a certainty, it will behoove these agencies to pool their resources and accomplish common goals in the psychological dimension.

Another major task facing PSYOP advocates after every election is that of educating the new players in DOD. This process should begin after a major turnover of senior civilian officials has taken place. A good vehicle to accomplish this is the DOD PSYOP Master Plan. Inclusion of key advisors in the final staffing of the document would provide an opportunity for them to educate themselves on psychological operations. There is some symbolic value attached to having the secretary of defense sign the master plan: It provides visible evidence of top-down emphasis.

Similarly, a number of DOD directives affecting military psychological operations need to be modified and revalidated. Involving key civilian players in this process would provide an excellent opportunity to educate them about PSYOP. The same holds true for staffing the defense planning guidance and the national military strategy, both of which should follow publication of the president's national security strategy. Indeed, the current national military strategy, published in January 1992, contains no mention of the psychological dimension.

Renewed efforts must be made to indoctrinate the mainstream of the officer corps on psychological operations. High-level emphasis, continuously applied, will be necessary to remedy the long-standing absence of PSYOP instruction in most of PME. Until this is done, OSD, JCS, and USSOCOM should coordinate their efforts and provide annual presentations to the senior service schools and the command and staff colleges. Formation of a traveling team to provide instruction on PSYOP to major commands should be considered. Examples of the successful use of psychological operations in Grenada, Panama, and the Persian Gulf would enhance the learning experience for commanders and staff officers. PSYOP should be integrated into computer-assisted war games and simulations. High-quality articles on psychological operations in professional military journals would help to publicize the successful use of this valuable force-multiplier. Institutionalization of military PSYOP will take root only when commanders are convinced that it is indispensable to combat effectiveness.

This education process must continue with planning for the utilization of military psychological operations. This will be a critical task during a period in which there is no apparent clear consensus about either US national security interests or the military's post-cold-war missions. Consider these words of Paul Hammond:

> Where strategy used to be something that we could think of as a plan to be executed, now it must be increasingly a set of capabilities for dealing with contingencies. Where strategy once was a matter of inventing the future, now it is increasingly a matter of adapting to it.[7]

This, of course, is what our present National Security Strategy and National Military Strategy attempt to provide: a broad focus, regionally oriented, and a "base force" to deal with an uncertain and rapidly changing world.

Within this context, military PSYOP advocates must demonstrate the applicability of psychological operations in all environments: support of conventional forces, low-intensity conflict, and special operations forces. Similarly, military PSYOP can play a valuable supporting role in the peacetime activities of the CINCs and other government agencies, such as the USIA and CIA.

First priority, however, must go to planning military PSYOP that will assist the unified commands—the combatant commands—and the conventional forces assigned to support them for various contingencies. As the military services reduce in size and units are withdrawn from Europe, a major realignment of Army PSYOP forces to other unified commands will take place. The CAPSTONE program provides an excellent vehicle to link these forces to the combatant commands for planning and participation in exercises. As the unified commands identify their priorities for contingency planning—a process of "emerging missions" that is likely to go on for some time—the PSYOP community must be alert to offer their services early on. In many cases, existing regional analytical studies will require revision, or the development of new ones, to provide the foundation for detailed PSYOP planning. Finally, supporting PSYOP plans should anticipate the new contingency plans of the unified commands and their assigned forces.

In addition to planning for mid- or high-intensity conflicts, the PSYOP community must plan for, and be prepared to participate in, peacetime and low-intensity conflict requirements. These requirements may include humanitarian assistance, peacekeeping, disaster relief, counterdrug operations, civic action, unscheduled studies and assessments oriented to crisis areas, advisory mobile training teams for the military forces of friendly nations, staff assistance to unified commands, counterterrorism, foreign internal defense, and support for DOD and non-DOD agencies. PSYOP forces—indeed, all special operations forces—provide versatile capabilities for the National Command Authority to consider in the uncertain future environment.[8]

Closely related is this major challenge: to continue modernization of the PSYOP force structure in terms of both personnel and equipment. For the PSYOP community, the corollary to Army Chief of Staff Gordon Sullivan's oft-quoted "no more Task Force Smiths" is "no more Tactical Information Detachments" (the only operational psychological warfare troop unit in the US Army when the Korean War erupted in June 1950; it consisted of two officers and approximately 20 enlisted men). In the face of continued pressures to reduce even more the size of the military, strenuous efforts must be made to avoid the fate experienced by the PSYOP community after every major conflict in this century. As recently as the early 1980s, the 4th Psychological Operations Group consistently reported the lowest readiness ratings in both personnel and equipment—and many reserve component units lacked items vital to mission accomplishment, including loudspeakers. This oversight must not recur.

In particular, the modest—but crucial—gains in personnel strength by the 4th Group during the revitalization of the 1980s must be preserved; the bulk of requirements for support of peacetime activities and low-intensity conflict will necessarily fall upon the active component. Over the long term, the key to military PSYOP viability will be the health of the Army's functional area for psychological operations (FA 39B). Continued high-level emphasis is needed to attract quality officers for this specialty, to provide them with excellent training, and to offer sufficient field grade billets for their utilization and career advancement.

The modernization of PSYOP-unique equipment undertaken in the 1980s, while beneficial, still has not been completed. The replacement for the TRT-22 radio, in particular, needs to be fielded. New loudspeakers are needed. Modernization, however, should be an ongoing process; emphasis needs to be placed on developing more mobile equipment for the future.

Being assigned to USSOCOM should help the PSYOP community in its perennial battle to obtain a proportionate share of dwindling resources or, put another way, to ensure that they do not suffer a disproportionate share of reductions.

However, a speedy resolution of whether or not military PSYOP will fully participate in MFP-11 for special operations forces is needed.

At the same time, the PSYOP community must demonstrate a fiscally responsible attitude by making maximum use of existing resources. This usage includes not only continued reliance on the reserve component for mobilization in mid-and high-intensity conflicts, but also in investigating ways in which specially trained reservists can be used to augment active component units for peacetime and low-intensity conflict requirements. Establishment of redundant capabilities should be avoided. For example, in the absence of a joint PSYOP center in the Washington, D.C., area, the 4th Psychological Operations Group's regional analysts should provide maximum support to the JCS and OSD PSYOP offices—in neither of which have area experts been assigned. The use of new technologies to enhance military psychological operations capabilities without increasing personnel requirements also must be pursued. Overall, the keys will be cost-effectiveness and the utility of military PSYOP in both peace and war.

Summary

US interest in military psychological operations during this century has been episodic, basically rising and falling during and after major conflicts. The Reagan administration interrupted this pattern by providing the impetus for a revitalization of US military PSYOP during peacetime. The international and domestic conditions that made this possible have changed significantly, however, and the PSYOP community faces formidable challenges in maintaining a viable capability for an uncertain future. To address these challenges, emphasis must be placed on educating senior civilian and military officials concerning the value of psychological operations, on aggressive planning to support the emerging contingency and peacetime missions of PSYOP within the unified commands, and on continued modernization of the PSYOP force structure.

After having prevailed in the battle to gain control of military psychological operations forces and policy, the commander in chief, USSOCOM, and the assistant secretary of defense for special operations and low-intensity conflict must play key roles in facing these challenges. For their part, the PSYOP community must demonstrate fiscal responsibility, maximum utilization of existing resources, and continued professionalism in accomplishing new missions. Working together, the goal of both the PSYOP community and its special operations leadership should be to ensure "no more Tactical Information Detachments" during this period of transition.

Notes

1. See Alfred H. Paddock, Jr., *U.S. Army Special Warfare: Its Origins* (Washington, D.C.: National Defense University Press, 1982), 12.

2. Ibid., 143.

3. Ibid.

4. Dennis P. Walko, "Psychological Operations in Panama (Just Cause and Promote Liberty)"; Frank L. Goldstein and Daniel W. Jacobowitz, "PSYOP in Desert Shield/Desert Storm," in US Special Operations Command unpublished manuscript "*Psychological Operations: Principles and Case Studies*" May 1992, 45.

5. Norman J. Ornstein, "Foreign Policy and the 1992 Election," *Foreign Affairs*, Summer 1992, 3f.

6. William J. Hyland, "The Case for Pragmatism," *Foreign Affairs*, American and the World, 1991–1992, 42f.

7. Paul Y. Hammond, "The Development of National Strategy in the Executive Branch: Overcoming Disincentives" in *Grand Strategy and the Decisionmaking Process*, ed. James C. Gaston (Washington, D.C.: National Defense University Press, 1992), 16.

8. For a more complete discussion of the versatility of SOF, see Carl W. Stiner, "U.S. Special Operations Forces: A Strategic Perspective," *Parameters*, Summer 1992, 2–13.

Blending Military and Civilian PSYOP Paradigms

Col Benjamin F. Findley, Jr., USAFR

This article compares and contrasts the concepts, philosophies, processes, strategies, and major elements of the civilian business marketing system (BMS) with the military PSYOP system. The purpose is to inform the reader of the many similarities and successful applications of proven BMS approaches, both macro and micro, and how they relate to strategic and tactical objectives in PSYOP scenarios. The goal is to improve the reader's understanding that integrating the effective practices of civilian marketing systems into the military PSYOP sector is very important. A related goal is to offer recommendations.

In a business firm, marketing generates revenues by persuading specific customers to do something. Whenever you try to persuade somebody to donate to the United Way, refrain from littering the highways, save energy, vote for your candidate, or buy your service, you are engaging in marketing. The challenge that faces the BMS is to generate those revenues and results by satisfying a targeted customer and making a profit in a socially responsible manner. Since every aspect of our lives is affected by business marketing practices, often in ways we do not even consider, we need to understand the basic principles and concepts of effective marketing. Then we need to apply them to PSYOP.

Business Marketing Philosophy

The American Marketing Association defines marketing as "the process of planning and executing the conception, pricing, promotion, and distribution of ideas, goods, and services to create exchanges that satisfy individual needs and organizational objectives." Inherent in this definition is an exchange process where two or more parties give something of value to one another to satisfy felt needs. The marketing function, simple and direct in subsistence-level economies, is more complicated in industrial societies. Nevertheless, the basic concept of exchange is the same. To understand the BMS philosophy, we must distinguish between a sales orientation and a marketing orientation. A sales orientation assumes that customers will resist purchasing products and services not deemed essential and that the task of selling is to convince them to buy something even if they do not need it. It is a "push" effort, focusing on selling after the product exists.

A marketing orientation is different—it begins with the customer and integrates marketing into each function of a business. Marketing will initially establish what the customer wants and needs; it should play a lead role in planning, coordinating, and directing the organization's activities. It certainly does involve selling, but only after the customer's need has been established. If the marketing job has been done well, the customer doesn't need much persuading.

Military Psychological Operations Philosophy

Military PSYOP is a specialized field of persuasive communications. It involves formulation, conceptualization, implementation, and evaluation of government-to-government and government-to-people persuasion. It is a planned use of human actions to influence the attitudes of populations that are important to national objectives. Lt Col Philip Katz, USA, Retired, emphasizes that the critical variable in military PSYOP is that of influencing the perceptions, and thereby the will, of foreign populations. PSYOP can modify the opinions, emotions, attitudes, and behaviors of a target audience. It can be directed at enemy forces to demoralize, disorient, and confuse them. It can induce defection, encourage dissident elements, and reduce combat effectiveness by adversely affecting the enemy's discipline and command and control functions. Directed toward neutral or friendly target audiences, it can be used to unite them, boost their morale, and provide information designed to foster their understanding and cooperation.

To enhance the total opportunity for success, PSYOP must be integrated into, and considered in, all military decisions. Certainly, PSYOP planning should be concurrent with—and coordinated with—all operational planning. A few commentators imply that there is some nefarious objective and some element of deliberate misinformation involved in PSYOP. Therefore, it is important that the public understand the truth: the United States government does *not* engage in public disinformation activities!

Business Marketing Process

A company's complete marketing system consists of many processes and variables, some controllable and some uncontrollable. *Marketing mix* is the term that describes the combination of the four main controllable inputs that constitute the core of a company's marketing process: (1) the product or service, (2) the promotional activities, (3) the distribution system, and (4) the price structure. Each input has its own process, with key decisions to be made and activities to be accomplished.

The product mix involves policies, procedures, services, and activities related to the product lines to be offered, the markets to sell to, and the degree of product differentiation. The product mix also involves decisions about which product quantities to offer with which brands, in which packages, where, and when.

The promotion mix process begins with an analysis of the organization's overall mission, from which specific promotion objectives are determined. The promotion process consists of decisions and activities related to advertising, personal selling, publicity, and sales promotion. Activities include analyzing the promotion's target audience; selecting appropriate themes and symbols; following market segmentation strategies; deciding on the optimal blend of print, broadcast, and other media; setting sales quotas; conducting message pretests; and analyzing selected characteristics of the target market. Motivation, perception, attitudes, opinions, and stress are considered. The promotion mix directly relates to the military PSYOP process and will be addressed later.

The distribution mix involves decisions about transportation, storage, and inventory. It also includes decisions about the type, quantity, and quality of wholesalers and retailers required.

The price mix includes methods for setting the right base price for a product. Discounts, allowances, markups, break-even points, and pricing strategy are elements of the price mix.

Aside from these four controllable elements in the marketing mix, the manager must be concerned with uncontrollable forces in the environment. Marketing is affected by cultural factors, laws, economic conditions, competition, technology, and demography, none of which are controllable; neither are the firm's microenvironment of suppliers, marketing intermediaries, and customers.

Military Psychological Operations Process

Psychological operations can be divided into three general categories: strategic, operational, and tactical. Strategic psychological operations are conducted to advance broad or long-term objectives designed to create a favorable global environment for military operations. Operational PSYOP seeks to achieve midterm objectives in support of regional campaigns and major theater operations. Tactical psychological operations are conducted to achieve short-term objectives against an enemy force, in direct support of tactical units. According to Field Manual 33-1, *Psychological Operations,* the mission of PSYOP forces is to support conventional as well as special operations forces at all levels of war and within all degrees of the conflict spectrum—low intensity, midintensity, and high intensity. Overlapping missions and objectives can blur the distinction between the categories of PSYOP.

Propaganda development for PSYOP involves a multistep process that is very similar to the promotion mix process for the BMS. Initially, the unit's mission must be analyzed so the military PSYOP mission can be derived. Information must be collected and the target must be analyzed. Just as in the business promotion process, the vulnerabilities and susceptibilities of the target audience are key factors. Appropriate themes, symbols, and media are selected, and propaganda messages are developed and pretested. Campaign approvals are acquired, propaganda is disseminated, and the impact of the psychological objective is assessed according to specific indicators.

Just as there are strategic, operational, and tactical levels of military PSYOP, there are corresponding levels in marketing planning. Louis E. Boone and David L. Kurtz define strategic marketing planning as "the process of determining the primary (long-term) objectives of an organization and the adoption of courses of action and the allocation of resources necessary to achieve those objectives." Tactical planning in marketing focuses on the implementation of short-term plans that must be completed near-term to implement overall strategies. Boone and Kurtz emphasize the necessity of adequate resource allocation in tactical planning. Certainly, this necessity should be recognized by military PSYOP planners.

Cultural Differences

The marketing manager in business and the military PSYOP specialist must understand a variety of influences on the promotion and persuasion processes in trying to satisfy the needs and wants of the target audience. Culture is at once the most basic and the broadest environmental determinant of individual behavior. Culture is comprised of values, ideas, attitudes, and other symbols that shape human behavior. The influence of culture is difficult to define because so many crucial elements are hidden from the participants of the system itself. Cultural patterns often exist as unrecognized assumptions that are taken for granted. Culture affects how and why people live and behave as they do, which affects the target audience's susceptibility to promotional themes and their buying/accepting behavior. America has many ethnic, religious, racial, and economic groups, each of which has its own cultural values. And each of these groups has sub-cultures with norms and values that are used to define meanings and to specify needs and wants. Thus, individuals respond uniquely to persuasive messages and appeals.

Persuasive messages in military PSYOP and in business marketing must use symbols and themes that are easily recognizable and meaningful to the culture they are trying to reach. A key persuasion guideline is that one should not substitute one's own judgment and cultural values for the values and cultural characteristics of the audience being targeted. No market has been more frustrating and difficult for American business marketers than the Japanese market. Japanese culture has produced marketing institutions and promotional relationships that differ from those found in the United States.

While leading some university students on an industrial tour of Japan in 1981, for example, I found that the food distribution system there was different from that in the United States. Japanese wholesalers, unlike their US counterparts, are specialized sellers. They do not carry a complete line of products for a select group of stores. Instead, they carry a select line of noncompetitive products that they supply to all supermarkets. Japanese wholesalers take great pride in supplying all retail stores with an exclusive but narrow product line. The goal of American wholesalers is total distribution of a complete product line to a captive set of stores; the goal of the Japanese wholesaler is distribution of a limited set of products to all retail outlets. Culture is a major influence on marketing and PSYOP.

Social Influences

The social groups to which people belong or aspire to belong help explain why they buy certain types and brands of products, shop in one type of store and not another, accept and believe certain ideas and persuasive messages, and spend their money in certain ways. A person's social class is a major influence on his style of life and an important factor in determining his social and economic behavior. A social class is a group of people whose members are nearly alike in terms of a characteristic that is valued in their group and that clearly differentiates them from others.

PSYOP specialists and promotion managers must be aware of social class influences on the acceptance of ideas and the willingness to behave a certain way. Different persuasive appeals, different copy, different art, and different media may be needed to promote the same product to different social classes. An upper-middle-class wife may want products, ideas, and brands that are clear symbols of her social status. She may be highly susceptible to preselling through mass media. Conversely, a lower-lower-class wife may buy largely on impulse and may be influenced by point-of-purchase promotion materials. She may not care to read much, and the broadcast media may be of great importance in communicating with her.

When social influences are combined with foreign culture influences, we see how difficult it is to understand needs and behaviors well enough to persuade an individual to accept a PSYOP message. One way to exert influence on a social class in a culture is to identify and target opinion leaders. They are chosen on the basis of expertise as well as social position. Exposure to mass media is significantly greater among opinion leaders than among nonleaders—and mass media directly influence opinion leaders. It is important to note that opinion leaders are far more effective in securing opinion changes among followers than are the mass media. Thus, it is important in PSYOP as well as in marketing to identify and aim persuasive messages directly at opinion leaders.

Family Influences

An individual is born into a family and inherits a certain social status. That status remains until he becomes an adult. Of all face-to-face groups, a person's family plays the strongest role in basic value and attitude formation.

Early family training is not cast off easily; it remains to affect one's individual values, ideas, and behaviors. Even as the adult individual strives to acquire the prestige symbols of another class, latent family influences remain to sway acceptance of ideas and products. Most buying decisions and idea acceptances are made within a framework of experience developed within the family.

Needs and Motives

Personal determinants of behavior include needs, wants, motives, and self-concepts. All of these variables combine with the previously presented interpersonal and group factors to influence one's acceptance of an idea or purchase of a product. People have a need when they lack something useful, required, or desired. When a few needs are fairly well satisfied, others emerge. Some say needs are insatiable, incapable of ever being fully satisfied; thus, PSYOP specialists and marketers do not have to satisfy one need before progressing on to other needs. Several needs at several levels can be present concurrently to influence an individual's behavior, but this approach complicates the prediction of behavior. Abraham H. Maslow proposed a hierarchy of needs that includes the most basic: physiological, safety, social, esteem; and the highest: self-actualization. Motives are inner states that direct people toward the goal of satisfying a felt need. Examples of motives include high quality, conformity, low price, performance, ease of use, long life, prestige, friendship, the opposite sex, and a desire to be different.

One's self-concept includes feelings, perceptions, and evaluations of oneself as a person. How one perceives and evaluates one's own status in relation to others in the same social class and reference groups—and to one's own levels of aspiration—are prime forces in shaping one's self-concept. The self-concept is learned; it emerges through experiences with the environment and through interactions with other people. By observing socially acceptable and unacceptable behavior and comparing one's own behavior to them, one lays the basis for one's own self-concept. The self-concept is useful in PSYOP and marketing because, through their actions and purchases of products, individuals let others know who they are and what they hope to be or to do. The purchase of goods and the expression of ideas serve as social symbols that communicate meaning. The behavior of an individual will be directed toward enhancing his self-concept through the consumption of goods and the expression of ideas as symbols, allowing a prediction of events.

Perceptions and Attitudes

Everything a person knows about the world and what it contains comes from perceptions. Perception is the reception of stimuli through the senses and the attachment of meaning to them. Sound, light, odor, taste, and pressure are received by one's sensory receptors—ears, eyes, nose, tongue, and skin. These sensations are translated into perceptions that are altered by previous learning, beliefs, values, and attitudes, and then organized into meaningful concepts. People perceive things that make sense within the context of their beliefs, values, attitudes, and experiences. Since we cannot possibly perceive all stimuli at any given time, our perceptions are highly selective and subjective; we see what we want to see and hear what we want to hear. Further, if we do not like what we perceive, we often distort or modify it.

Mass-marketing messages and PSYOP messages are based on selective perception. Because a large number of persuasive messages are directed at different target audiences, PSYOP specialists and marketers must carefully test the messages to be employed with representative targets. They must make every effort to ensure that the various target groups perceive the messages that were intended.

Attitudes are predispositions to respond in particular ways toward people, ideas, activities, or things. An attitude is not neutral; rather, it is for or against some person, object, or idea. An attitude is composed of three dimensions: cognitive (factual information), affective (feelings and emotions), and behavioral (tendencies to behave in a certain way). Attitudes affect both perception and behavior, and people resist changes in their attitudes. The more closely an attitude is related to one's self-concept, the greater will be one's resistance to changing it.

To change an attitude, a PSYOP specialist must (1) provide new information to enlarge and change the cognitive dimension; (2) attack the affective dimension of the attitude by associating the end-state of change with the desirable consequences that result; and/or (3) induce the person to engage in "attitude-discrepant" behavior (behavior that contradicts held preferences). Successful attitude-change approaches include using a trustworthy and credible source, drawing conclusions for the target, repeating the message, and using both one-sided and two-sided messages.

Successful and Unsuccessful Business Marketing

Businesses spend several billions of dollars each year on marketing, a fact that indicates the importance they place on it. According to an *Advertising Age* article, the top 100 firms spent more than $34 billion in 1989 for advertising. Advertising represents about 20 percent of total marketing expenditures. The top 10 advertisers spent more than $10 billion for advertising alone in 1989. Those companies have been successful. Failures do occur, however. A 1988 *Advertising Age* report, for example, lists these $100 million failures: Campbell's Red Kettle Soups, Ford's Edsel, Du Pont's corfam, Hunt's flavored ketchups, and Gillette's Nine Flags Cologne.

Newspapers account for the largest annual advertising expenditure ($31.2 million in 1988). Other 1988 expenditures: television ($26.1), direct mail ($21.2), radio ($7.7), and magazines ($6.1). Use of advertising, however, does not guarantee success even if it is planned and implemented well. A key reason is that most people do not systematically process the information they receive through the marketing message. We are largely emotional, influenced by testimonials and emotional appeals—we do not logically consider pertinent criteria when we purchase products. It seems we are more influenced by the skill, appearance, and character of the presenter than the content of the message. A 1988 *Advertising Age* survey mentioned three of the most powerful ads in US history: Volkswagen's "Lemon," Marlboro's "Country," and Coca-Cola's "Real Thing." Aside from getting commitments from dealers to carry the product, the Coca-Cola ad was successful because it capitalized on the affective domain and the unsystematic emotional responses of people. A critical conclusion for PSYOP specialists and marketers is that the ad created a psychological idea of a unique and better taste. It focused on image and emotional themes through heavy repeat advertising and readily identifiable names, symbols, and packages. It also linked abstract, beneficial images to the physical product—the implied associations of Coca-Cola with youth and authenticity.

Business Marketing Models

Marketing lessons can be learned from business failures as well as from effective marketing models. Burger Chef did not realize that economic conditions are major factors, did not adapt to changing consumer preferences, and did not meet target market expectations. Coors did not recognize changes in the environment that necessitated adjustments in marketing strategies, and did not take the time to continue developing an image of quality and great desirability. The World Football League lost credibility by taking actions counter to prior promises and by not paying strict attention to environmental variables.

After analyzing five leading models of consumer behavior, I have blended optimal elements from each to suggest a marketing model for the buyer's decision process. The model focuses on systematic thinking and defining relevant elements for making a buying decision. It attempts to explain the progressive process that a consumer might experience in buying products or accepting ideas. PSYOP specialists and marketers can anticipate effective messages if they understand the process through which buyers can progress. They can then address those messages to bring about the desired change in behavior. My buyer decision model is like any scientific problem-solving model except that it focuses on buyer behavior. It consists of five stages: problem recognition, information gathering, alternative evaluation, the decision to buy, and postpurchase evaluation. Initially, the target consumer recognizes a problem because of inadequate supplies, changing needs, or other internal or external activities. This first stage forces buyers to succinctly state the problem based on specific unfilled needs. Then the second stage begins the planned search for specific, relevant, and identical information on all possible products. Buyers must weigh the pros and cons of the various products based on a set of criteria during this third stage. In the next stage, they make the decision that best meets their needs and wants, and that satisfies the expectations of influencers in their social groups. The fifth and final stage is the evaluation of that decision; it should not be omitted.

Effective Objectives

Specific marketing and promotion-persuasion objectives, based on the organization's general goals, should be established. Clearly stated objectives should exist for each promotion tool so that results can be measured against them. Advertising, personal selling, publicity, and sales promotion functions should have specific expected objectives that are measurable. Print, broadcast, and audiovisual media should also have specific objectives. Marketing objectives should meet certain criteria and should be built on a solid foundation of research that reveals what the targets want to buy. The objectives should be stated in terms of target benefits—and in quantitatively measurable terms. They should help attain and reinforce the firm's overall objectives, and they should be reviewed periodically.

The Ethics of Persuading

Many critics of marketing contend that persuasion is really manipulation—or even control—of consumers, unethically coercing them into buying products they do not want. Is it immoral to arouse fears of inadequacy to sell an antiperspirant? Is it wrong to use reference-group concepts to make a product more appealing? Is it improper to design products and appeals that enhance a buyer's self-concept in order to gain a sale? Some observers argue that embellishment and distortion are legitimate and socially desirable activities and that illegitimacy enters the picture only if there is falsification with larcenous intent.

That argument notwithstanding, however, truth is the best persuasion approach; it should be the fundamental principle in both business marketing and military PSYOP. Truth, which is basic to strategic peacetime foreign policy, should be our fundamental guideline. The ultimate questions concern "What is the truth?" and "Can consumers really be convinced to buy something they do not want?"

Target Market Analysis

Target market analysis is performed every time a marketing or PSYOP campaign is developed. It includes selecting a target, determining conditions that affect the target, analyzing target vulnerabilities, specifying target susceptibilities, formulating the persuasive objective, determining target effectiveness, and assessing campaign impact indicators. Once the conditions and vulnerabilities of the Iraqi soldiers were identified in the Persian Gulf War, individual susceptibilities for each vulnerability were specified. Because the coalition's bombing campaign was effective, the Iraqis had needs for food, shelter, water, medical attention, and basic safety. They were, therefore, driven to respond in an expected way. Target market analysis is key to successful marketing and PSYOP.

Recommendations

My recommendations, which are based on my own analysis, are intended to serve as suggestions for improved business marketing and military psychological operations. They emphasize the transfer of proven marketing and promotion strategies, concepts, and techniques from the civilian business marketing system to the military PSYOP system.

• A primary recommendation is to accept the marketing orientation (as opposed to a sales orientation). The marketing orientation begins with the customer's needs and wants, then integrates marketing into each function of the business to design the desired product. (The sales orientation focuses on the product characteristics and attempts to sell the existing product.)

Both PSYOP planning and marketing planning should be concurrent with, and coordinated with, all operational planning. Together, they should address the target audience's needs in any given situation. This relates to the function of PSYOP as a force multiplier in all types of military activity.

• Another recommendation is to ensure that marketing and PSYOP support *all* operations, to include production and finance in business or conventional and special operations in the military. This support should be for all levels of business operations and all levels of war—strategic, operational, and tactical; low intensity, midintensity, and high intensity.

• A commensurate resource and budget commitment to support adequate PSYOP and marketing is required so that PSYOP and marketing can be integrated into all operational planning. It is an absolute necessity that military PSYOP and business marketing planners understand the level of resourcing necessary to conduct a successful campaign; undercapitalization may doom an otherwise promising organization to failure.

• The business community and the military services must accept culture as the most basic environmental determinant of individual behavior, and they must gather sufficient quantitative and qualitative information about it. Individual judgment and cultural values should not be substituted for those of your target audience.

• PSYOP specialists and marketing managers must be aware that social-class membership influences the acceptance of ideas, the willingness to behave a certain way, and the need to use different persuasive techniques and concepts.

• Persuasive messages should be aimed directly at opinion leaders; they are far more effective in securing changes among followers than are the mass media.

• It is important to make maximum utilization of family influence; research shows that it plays the strongest role in basic value and attitude formation.

• The PSYOP specialist and the business marketer must know the various needs of individuals and the corresponding specific persuasive appeals that can be employed to influence each need.

• Marketing and PSYOP professionals must recognize the dominant influence of the self-concept—and how consumption of goods and expression of ideas are directly based on that concept and on selective perception. They must understand the main behavioral dimensions of attitudes, so that appropriate techniques can be employed to change them.

• My final recommendation is an all-encompassing one. It refers to the need to educate our military and civilian leaders about marketing philosophy and psychological operations (as a unique weapons system). Also, managers and specialists already employed in the field need tactical training. The professional military education service school system should include courses designed to improve the awareness and understanding of PSYOP. The primary goal would be to enhance the understanding of strategic PSYOP and marketing, their benefits and long-term capabilities, and their applications in support of our national objectives.

Bibliography

Air Force Field Manual (FM) 2-5. "Psychological Operations." *Tactical Air Operations, Special Air Warfare.* Washington, D.C.: Government Printing Office (GPO), 1967.

Army FM 33-1. *Psychological Operations.* Washington, D.C.: GPO, 1987.

Black, M.A. "Tactical Psychological Operations." *Canadian Defense Quarterly.* Summer 1987, 30–33.

Boone, Louis E., and David L. Kurtz. *Contemporary Marketing.* Chicago, Ill.: Dryden Press, 1989.

Bovee, Courtland L., and William Arens. *Advertising.* Homewood, Ill.: R. D. Irwin, 1982.

Clausewitz, Carl von. *On War.* Edited and translated by Michael Howard and Peter Paret. Princeton, N.J.: Princeton University Press, 1976.

Crane, Barry, et al. "Between Peace and War: Comprehending Low-Intensity Conflict." *Special Warfare.* Summer 1989, 4–25.

Cunningham, William H., et al. *Marketing: A Managerial Approach.* Cincinnati, Ohio: South-Western Publishing Co., 1990.

Department of the Army. *The Art and Science of Psychological Operations: Case Studies of Military Applications,* vol. 1. Washington, D.C.: Institute for Research in the Behavioral Sciences, 1976.

Department of Defense. Headquarters 2d PSYOP Group, US Army. *Psychological Operations.* Washington, D.C.: GPO, 1989.

Findley, Benjamin F. "USSOCOM's Foreign Internal Defense Mission Is Real and Critical." *United States Special Operations Command Update.* Spring 1990, 6–7.

Garth, S. Jowett, and Victoria O'Donnell. *Propaganda and Persuasion.* Newbury Park, Calif.: Sage Publications, 1986.

Goldstein, Frank L. "Psychological Operations, Terrorism and Glasnost." *United States Special Operations Command Update.* Spring 1990, 18–19.

Gordon, Joseph S. *Psychological Operations: The Soviet Challenge.* Boulder, Colo.: Westview Press, 1988.

Hart, B.H. Liddell. *Strategy.* New York: Frederick Praeger Publishers, 1968.

Joint Chiefs of Staff. *Dictionary of Military and Associated Terms.* Washington, D.C.: GPO, 1984.

Katz, Philip. "PSYOP and Communications Theory," in *The Art and Science of Psychological Operations: Case Studies of Military Application.* Washington, D.C.: Department of the Army, 1976.

_____. "Tactical PSYOP in Support of Combat Operations," in Ron D. McLaurin *Military Propaganda Psychological Warfare and Operations.* New York: Frederick Praeger Publishers, 1982.

Kaufman, Louis. *Essentials of Advertising.* New York: Harcourt Brace Jovanovich, Inc., 1980.

Kriesel, Melvin E. "The Use of Psychological Operations In Counterinsurgency." Langley AFB, Va.: Army-Air Force Center for Low Intensity Conflict, November 1988.

Lord, Carnes, and Frank E. Barnett. *Political Warfare and Psychological Operations.* Washington, D.C.: National Defense University Press (NDUP), 1989.

McCarthy, Jerome, and William Perreault. *Basic Marketing.* Homewood, Ill.: R. D. Irwin, Inc., 1987.

McLaughlin, Mark W. "The Future of Psychological Operations." *Perspectives.* Spring 1991.

McLaurin, Ronald D. "Perceptions, Persuasion, and Power." *Military Propaganda: Psychological Warfare and Operations.* New York: Frederick Praeger Publishers, 1982.

Office of the Secretary of Defense. *Psychological Operations Master Plan.* Washington, D.C.: GPO, 1990.

Orth, Richard. "Source Factors in Persuasion." *The Art and Science of Psychological Operations: Case Studies of Military Application.* Washington, D.C.: Department of the Army, 1976.

Paddock, Alfred H. "Military Psychological Operations and U.S. Strategy," in J.S. Gordon, ed., *Psychological Operations: The Soviet Challenge.* Boulder, Colo.: Westview Press, 1988.

_____. *U.S. Army Special Warfare: Its Origin.* Washington, D.C.: NDUP, 1982.

Pedersen, Judith. *Fundamentals of Modern Marketing.* New York: American Management Association, 1978.

Pressley, Milton M. *Marketing Planning*. New Orleans, La.: Marketing Management and Research Institute, 1978.

Reppert, John C. "Psychological Operations: The Oldest Weapon of Mass Destruction." *Special Warfare* 1, no. 2 (July 1988): 28–33.

Resource Book for Psychological Operations. Hurlburt Field, Fla.: United States Air Force Special Operations School, 1988.

Romerstein, Herbert. *Psychological Operations and Political Warfare in Long Term Strategic Planning*. New York: Frederick Praeger Publishers, 1990.

Runyon, Kenneth E. *Consumer Behavior*. Columbus, Ohio: Merrill Publishing, 1987.

Sconyers, Ronald T. "The Information War." *Military Review*. February 1989, 44–52.

Stanley, Richard E. *Promotion*. Englewood Cliffs, N.J.: Prentice-Hall, Inc., 1982.

Stanton, William J. *Fundamentals of Marketing*. New York: McGraw-Hill Book Company, 1987.

Sun Tzu. *The Art of War*. Translated by Samuel B. Griffith. Oxford: Oxford University Press, 1963.

Volkogonov, D. A. *The Psychological War*. Moscow: Progress Press, 1986.

Walker, Fred W. "Strategic Concepts for Military Psychological Operations." Joint Chiefs of Staff Unpublished Report, Washington, D.C., 1987.

_____. "Truth Is the Best Propaganda." *National Guard*. October 1987, 27.

Yudovich, Lev. "Assessment of the Vulnerability of the Soviet Soldier." Fort Bragg, N.C.: John F. Kennedy Special Warfare Center and School, 1987.

PART II

National Policy
and
PSYOP Planning

Introduction

This section presents articles about recurrent psychological issues and roles in formulating and implementing our national policy, objectives, and strategy. The authors explain historical and contemporary elements of the national policy process in PSYOP and the framework within which national PSYOP policy is formulated, administered, and implemented.

Dr Carnes Lord explores the historical influences on developing effective US strategy, doctrine, and organizational structure for the conduct of psychological and political warfare. While accepting that a psychological-political component is inherent in every use of the diplomatic, economic, and military instruments of national power, he urges a rethinking of the role of political and psychological factors and a revitalization and integration of their capabilities into US national strategy. Dr Lord analyzes the influences of American cultural inhibitions and the media on psychological conflict.

Michael A. Morris reminds us that national policy objectives must be clearly defined so that military objectives and strategy can be precise. He emphasizes that clarification of political objectives is the responsibility of political leaders rather than military leaders. Morris concludes that propaganda advantages can be gained through clearly stated political objectives. He suggests guidelines for distinguishing clear political objectives from vague ones.

William F. Johnston focuses attention on psychological operations as a vital instrument in national liberation wars. He believes governments of countries threatened with insurgency should regard PSYOP, particularly face-to-face communications, as a first line of internal defense—the Vietnam lesson. He also wants more recognition of PSYOP as a major instrument of low-intensity conflict and the elevation of PSYOP planners to "first-team" status.

Lloyd A. Free examines the role of public opinion and the psychological dimension in international security affairs. He believes that while public opinion cannot be slavishly followed, psychological data should be collected and analyzed so that government can take this factor into account in planning.

Lt Col Philip K. Katz, USA, Retired, Ronald D. McLaurin, and Preston S. Abbott include a critical analysis of US PSYOP. They focus on the communications aspect of international relations and PSYOP, identify and assess the principles and critical developments in this field, and advance ideas for progress. They stress that intelligence is the most commonly overlooked prerequisite to effective PSYOP. Their focus on evaluating the effectiveness of PSYOP challenges the reader to systematically concentrate on results. This article is clearly a historical piece that sets the groundwork for producing effective PSYOP.

The Psychological Dimension in National Strategy

Dr Carnes Lord

To recall the time when psychological-political warfare was widely acknowledged by Americans as an important instrument of national strategy requires a certain effort of historical imagination. Such was indeed the case, however, from the early days of the Second World War until the mid-1960s. The experience of World War II convinced many American political and military leaders that the psychological dimension of conflict had become critical in the contemporary world.

Totalitarian regimes that employed modern communications technologies to specialize in ideology and subversion constituted a qualitatively new strategic problem for the West. The postwar years witnessed an outpouring of academic studies in this area, most of which took for granted the necessity and legitimacy of a vigorous American response to the emerging political-ideological threat posed by the Soviet Union and the international communist movement.[1]

The US government allowed its propaganda and political warfare capabilities to wither in the years of rapid demobilization immediately following World War II. Creation of the Central Intelligence Agency (CIA) in 1947, however, provided a fresh impetus and an organizational vehicle for covert psychological operations and political action in peacetime. The deteriorating political situation in Western Europe in the late 1940s created urgent objectives for the CIA's covert action directorate.

The coming of the Korean War stimulated improvements in US overt capabilities as well. In 1950, President Harry S Truman created the Psychological Strategy Board in the White House to provide a high-level focus for government-wide

73

activities in this area. The new International Information Administration was established within the State Department; military psychological operations were given new life (the Army established the Psychological Warfare Center at Fort Bragg, North Carolina, in 1952); and the Psychological Operations Coordinating Committee attempted to provide operational coordination among the various involved agencies.[2] President Dwight D. Eisenhower moved to create the US Information Agency (USIA) as an autonomous agency reporting directly to the National Security Council.[3]

It can be questioned whether the US government as a whole was able, in the 1950s, to develop an effective doctrine and organizational structure for the conduct of psychological and political warfare. Sharp differences of opinion existed over fundamental questions of strategy toward the Soviet bloc and over the roles of key agencies such as CIA and USIA—and coordination among the agencies left much to be desired.[4]

It could perhaps be argued that militarization of the Vietnam War was the key factor underlying the progressive atrophy of US political warfare capabilities after the mid-1960s. However, it seems evident that larger issues of governmental organization and national style or culture also figured critically in this development. It should have been clear that the assignment of political warfare responsibilities to new agencies would not eliminate—and in certain respects would probably fortify—the sources of bureaucratic resistance to this new instrument of national power.[5] Finally, mention must be made of the cultural revolution that took place in the United States beginning in the mid-1960s. The shattering of the foreign policy consensus of the postwar decades (a result of Vietnam) meant, in the first place, a questioning of American values and the legitimacy of a leading world role for the United States. Such attitudes could only spell trouble for any strategy that depended on the confident projection abroad of America's political identity and values.

Any attempt to rethink the role of political and psychological warfare in US strategy today must take account of these fundamental and persisting obstacles. Before turning to consider them more systematically, however, it is necessary to sketch briefly the basic features of political and psychological warfare and their relationships to other instruments of national power.

One problem is the general tendency to use the terms *psychological warfare* and *political warfare* interchangeably, not to mention a variety of similar terms—*ideological warfare, the war of ideas, political communication, psychological operations,* and more. The uncertainty of reference derives partly from the fact that this sort of warfare is waged to a considerable extent with weapons that are not truly distinctive. There are indeed distinctive psychological instruments—radio broadcasting, publications of various kinds, and educational and cultural programs—that are capable of communicating information and ideas. But because these capabilities are easier to conceptualize, and to handle bureaucratically, the tendency has been to give them undue weight when it comes to defining the overall phenomenon.

There is a psychological dimension to the employment of every instrument of national power, emphatically including military force at every level. Similarly, major increments of military and economic power necessarily generate political effects. In thinking about psychological and political warfare, the tendency has been to think about the conflict of ideas, ideologies, and opinions. Yet the concept is in fact seriously misleading. Psychological-political warfare is also about cultural and political symbols, perceptions and emotions, behavior of individuals and groups under stress, and cohesion of organizations and alliances.

Using the term *warfare* to describe US psychological-political strategy in its broadest sense is itself problematic. Psychological-political operations need not be directed only to adversaries; neutral, allied, and semiallied nations potentially constitute highly important targets because our enemies will target them in efforts to break those alliances.

The English language apparently does not include a good term for designating psychological-political operations in their broadest sense. In recent years, and especially since the arrival of the Reagan administration, the term *public diplomacy* has gained considerable currency in Washington. Public diplomacy appears to encompass three distinct though closely related functions: international information, international political action (or what may be called overt political warfare), and public affairs. The inclusion of public affairs is a recognition that it is impossible in a modern democracy to separate sharply information communicated to domestic audiences from that communicated to international audiences; but the domestic function associated with public diplomacy differs from traditional public affairs by virtue of its strategic approach and its active effort to shape the domestic political agenda.[6]

The public diplomacy rubric serves a useful purpose, but it is not and was not intended to be comprehensive. Covert political warfare was excluded from its purview from the beginning. Nor does it have any clear relationship to military psychological operations, to educational and cultural affairs, or to the range of US government activities that may be grouped under the general label of international aid and humanitarian affairs.

I would propose, if only for the sake of clarity, the following anatomy of basic psychological-political warfare functions.

Political warfare is a general category of activities encompassing political action, coercive diplomacy, and covert political warfare. In general, the first of these functions is performed by diplomatic personnel, the second by military and diplomatic personnel, and the third by intelligence personnel. *Political action* refers to a range of activities, including certain kinds of multilateral diplomacy, support for foreign political parties or forces, and support for or work with international associations of various kinds.[7] *Coercive diplomacy* refers to diplomacy presupposing the use or threatened use of military force to achieve political objectives.[8] *Covert political warfare* corresponds roughly to the covert aspects of what the Soviets called active measures; it includes support for insurgencies, operations against enemy alliances, influence operations, and black propaganda.[9]

Psychological operations, once frequently used in a general sense to designate psychological-political operations as a whole, is probably best reserved for use as a term of art to designate military PSYOP.[10] Military PSYOP can encompass both overt and covert activities in both peacetime and war, and its scope can vary from the tactical battlefield to the operational and strategic levels of conflict. Historically, however, US military interest in PSYOP has focused heavily on tactical applications in wartime. Battlefield PSYOP is sometimes distinguished from consolidation PSYOP, which is geared to securing the loyalty and cooperation of civilian populations in combat areas and is closely related to civic action conducted by military forces in low-intensity conflict situations. Another related function is troop information or education, which serves among other things—much like public affairs in relation to public diplomacy—to counter the psychological operations of the enemy.

International communications encompasses international information and international educational and cultural affairs. USIA performs this range of functions, though other organizations—in particular, "Radio Free Europe/Radio Liberty" but also the Departments of State and Defense—perform information functions of political or strategic importance as well.[11]

A further general category is *international aid and humanitarian affairs.* This category includes foreign economic and development aid, food aid, humanitarian assistance (rescue operations, disaster relief, famine relief, etc.), and technical assistance of various kinds. Many agencies involved in such activities, including the Defense Department, the Agency for International Development, and the Peace Corps, are organizations that have dedicated missions in this area. Although these functions are bureaucratically scattered and very largely autonomous, they have a very important psychological-political component. Whether intentionally or otherwise, they serve as significant instruments of US foreign policy and national strategy.

Finally, to repeat what was said earlier, a psychological-political component is inherent in every use of the diplomatic, economic, and military instruments of national power. The art of negotiation rests on an understanding of individual and group psychology and a sensitivity to cultural contexts. Similarly, the exercise of military command at all levels involves an assessment of the psychological strengths and vulnerabilities of the enemy commander and his forces; deception and surprise are key elements of the military art. A

77

nation's economic and military strength creates political weight that can be exploited in a variety of ways to advance the national interest.

The impetus for rethinking the role of political and psychological factors in US national strategy has come primarily from the renewed attention given these matters by the Reagan administration.[12] Since 1981, a major effort has been under way to modernize and expand US government capabilities in the area of international communications, particularly radio and television broadcasting.

It remains doubtful that the administration has succeeded in overcoming the obstacles, both internal and external, to a thorough revitalization of US psychological-political capabilities and their full integration into national strategy.[13] In many respects, the cultural pressures working against such an effort are as strong as—or stronger than—ever. In addition to a kind of generic resistance to such activities on the part of Americans as Americans, there has been a wholesale loss of understanding and support of them among American elites in recent years. But perhaps equally troublesome is the resistance stemming from the national security bureaucracy itself and from the continuing weakness of integrated strategic planning and decision making at the national level.

Painful as it may be to face squarely the question of American cultural inhibitions in the area of psychological-political conflict, the effort is necessary—in order not only to develop intelligent approaches for dealing with them but also to achieve cultural self-consciousness, which is essential for effective participation in this kind of conflict. It is essential because Americans tend to assume that people everywhere are much like themselves, with similar fundamental motivations and views of the world. But blindness to differences in national characteristics is apt to be a fatal handicap for anyone attempting to affect the psychological orientation and political behavior of foreign audiences.

Perhaps the most severe single limitation in the American outlook is its tendency to discount the relevance of nonmaterial factors such as history, culture, and ideas. Americans tend to assume that concrete interests such as economic well-being, personal freedom, and security of life and limb are the critical determinants of political behavior everywhere. It is an interesting irony that such a view is so prevalent in a country as fundamentally idealistic as the United States while the importance the Soviets attributed to ideological factors stands in some tension with the materialistic basis of Marxism.

Connected with this emphasis on material considerations is the fact that Americans, unlike many peoples, are uncomfortable with personal confrontation and argument and do not customarily debate political and ideological questions in their private lives. Americans tend to look on the political realm as an arena not of conflict and struggle but of bargaining and consensus, where strongly held opinions and principled positions are disruptive of the process and to be discouraged. This tendency makes it extremely difficult for Americans to deal effectively in international settings where basic American values are under challenge. Furthermore, American notions of fair play and due process are subject to serious misinterpretation abroad. Americans' insistence on a presentation of both sides of any argument is frequently seen as reflecting a lack of self-confidence. In general, the openness and penchant for self-criticism in American society strike many foreigners as manifestations of weakness rather than strength.

Manifest or latent in the attitudes of many Americans toward the practice of psychological-political warfare is a distaste for any sort of psychological manipulation or deception. The idea that psychological-political warfare is a black art that can be morally justified only under the most extreme circumstances is a derivative of such attitudes. That such activities necessarily involve misrepresentation or deception is in any case far from the truth. (The conveying of purely factual information under certain circumstances can have powerful psychological effects.)

Military psychological operations, such as battlefield broadcasting, have as their primary purpose the saving of lives—enemy as well as friendly lives. Indeed, such activities make both moral and strategic sense. According to the Chinese strategist Sun Tzu, "what is of supreme importance in war is to attack the enemy's strategy. . . . Next best is to disrupt his alliances. . . . The next best is to attack his army. . . .

The worst policy is to attack cities." As Sun Tzu puts it, "To subdue the enemy without fighting is the acme of skill." In other words, competence in the psychological-political sphere is the essence of a rational approach to war.[14] Failure to attain such competence within the limits of one's possibilities is a failure that is all too likely to be paid for in blood.

As important as the effect of these general cultural biases is the role of the American media. Developments in the culture and operating style of the prestige media in the United States in recent years have substantially complicated any effort by the US government to engage seriously in psychological-political conflict. Before the late 1960s, it may be argued, a satisfactory understanding existed between journalists on the one hand and American military and government officials on the other regarding the proper scope and limits of press coverage of national security and foreign policy matters. In particular, the press in wartime tended to adopt the national cause and to accept broad responsibility not only for protecting sensitive information but also for safeguarding the morale both of the troops at the front and of civilians at home

Since Vietnam, there has been a dramatic change.[15] In the general wreck of the national foreign policy consensus resulting from that experience, the media have adopted an increasingly skeptical attitude—not only toward the specific policies and actions of the incumbent administration but also toward many of the fundamental assumptions that had underpinned the global position and role of the United States since World War II. The legitimacy of the American defense and intelligence establishments in particular has been sharply questioned and subjected to scrutiny by the new style of investigative journalism inaugurated by the prestige press. Most significantly, the media ended their deference to and informal cooperation with an incumbent administration in favor of a posture of neutral observer or critic. One result of this shift has been a general refusal to take responsibility for the consequences of media coverage's effect on national security policy outcomes.[16]

This change in media attitudes is worth dwelling on, since the role of the media on the battlefield of the future is likely to decide whether the United States will be capable of conducting effective military psychological operations. In general, the media now acknowledge a responsibility to avoid jeopardizing the lives of American soldiers engaged in military operations. But they do not recognize an obligation to refrain from publicizing information that demoralizes American troops, reveals aspects of American intelligence or military planning, undermines American diplomatic initiatives, or gives psychological aid and comfort to the enemy. This obligation is denied even though the ultimate effect of such disclosure may be to prolong military operations and cost American lives, not to speak of more generally damaging the international position of the United States and its ability to avoid future conflicts. Nor do the media recognize any obligations with respect to the domestic audience.

Of particular importance in this connection is the wartime role of television. To argue (as media spokesmen regularly do) that television coverage is essential to informed debate on the merits of a particular military action is unconvincing, not to say disingenuous. The information content of TV pictures is typically low or nonexistent, and the emotions such pictures arouse are more likely to defeat than to promote rational discussion. The rapid juxtaposition of death and destruction images torn out of any intelligible context, so common in television coverage of war, inevitably encourages the feeling that the war is futile, immoral, or absurd.[17]

Equally harmful is the practice—pursued well beyond the point of abuse by the networks in Lebanon in 1983—of interviewing American GIs on their feelings and views about the situation they happen to be involved in. To portray soldiers (and if they are looked for they will be found) who are confused, inarticulate, naive, or bitter about the reasons why a war is being fought or the way it is being conducted serves no purpose. The immediate danger to morale and the effect on allied and enemy perceptions are only part of the costs of such behavior. As in the case of media obsession with the families of terrorist victims, the effect is to pander to private concerns and emotions and to mobilize them in a way that greatly complicates the pursuit of rational policies by the US government.

All of this suggests that serious thought needs to be given to restricting or even eliminating at least television's presence on the battlefield of the future, with or without the cooperation of the media. Particularly difficult, of course, is the question of censorship or restraint of the media during limited contingencies or undeclared wars such as Vietnam, Lebanon, or Grenada. Because the stakes in such conflicts are relatively low, the pressures for preserving peacetime rules of media engagement are difficult to resist. Yet it is precisely these conflicts in which the political and psychological element in war is predominant. These conflicts are therefore most directly susceptible to influence by media reporting. Devising acceptable arrangements for limiting media coverage of such wars in the future may well be critical if the United States is ever to engage in them successfully.[18]

The American media have also affected the US government's international information programs. Despite the popular image of the "Voice of America" and "Radio Free Europe/Radio Liberty" as propaganda organs fully comparable to "Radio Moscow," they have been profoundly affected by the American media's evolution over the last two decades, as well as by the general cultural climate these media have reflected. Objectivity and balance as understood by the new journalism have become the standards for these radios as well. The point here is not that a balanced treatment of American or Soviet virtues and vices is not in some sense desirable, but rather that the domestic cultural context shapes—to an unhealthy degree—the aims and methods of the US international radios.

The effective conduct of psychological-political warfare by the United States is perhaps more immediately constrained by bureaucratic and organizational weaknesses within the US government itself. There is a connection between the inadequacy of US psychological-political warfare efforts in the past and the inadequacy of strategic planning and decision making at the national level. Precisely because the instruments of psychological-political conflict are not altogether distinctive, this arena requires fully integrated planning and coordinated operations throughout virtually the entire national security bureaucracy.

We need not dwell at length on the causes of the resistance to psychological-political warfare throughout the US diplomatic, military, and intelligence establishments; they are apparent to most of those who have had direct experience in this field. The State Department continues to ply its trade very much in the spirit of the foreign ministries of nineteenth-century Europe, with only grudging accommodation to the role played by modern communications, public opinion, ideology, and political theater in contemporary international affairs.[19] The military services—in their preoccupation with technology, major weapon systems, and the big war—tend to neglect low-cost approaches to enhancing operational effectiveness, especially at the lower end of the conflict spectrum; and they tend to regard psychological-political warfare as someone else's business.[20]

The failure of both the State Department and the military to assume psychological-political responsibilities might seem to point to the intelligence community as the natural home for such activities. Yet the CIA has generally been unwilling or unable to allow the degree of coordination with other governmental entities that is essential for an integrated national strategy in this area. There is also a feeling at the agency that the era of CIA involvement in political or ideological struggles is essentially past—that an aggressive agency role can only jeopardize more important institutional equities.

What, then, would be involved in a revitalization of US psychological-political warfare capabilities? The foregoing discussion is not meant to suggest that fundamental change is a hopeless proposition—only that it is essential for us to be conscious of the obstacles to it. We must be particularly conscious of the less tangible obstacles that cannot be fixed by organizational rewiring or other short-term measures. At the same time, useful steps undoubtedly can be taken without creating a revolution in the way Americans behave or the US government conducts its business.

Perhaps the most promising area for change is in the field of military psychological operations. The fact that military psychological operations have generally been treated as a subspecialty of special operations is a good indication of the conceptual and operational limitations under which it has long labored.[21] Of course, the very identification of PSYOP as a special forces mission has tended to isolate it from normal military activities and bring it under a certain suspicion, which its black connotations have further strengthened. But PSYOP has perhaps suffered most from identification with the hardware and missions of the tactical battlefield—that is, leaflet delivery, loudspeakers, and radio broadcasting. As a result of all this, PSYOP has had very low priority in terms of personnel, equipment, training, exercising, and doctrine. In addition, it has suffered from low visibility at senior command levels within the military (particularly outside the Army, which owns most PSYOP assets), not to speak of other US government organizations.

This situation is now beginning to change as a result of renewed interest within the military as well as at the national level. The rethinking of PSYOP roles and missions is still at an early stage, however, and basic doctrinal and organizational questions remain to be worked out. The Air Force and Navy appear not yet fully persuaded that PSYOP can be a responsibility of all the services and of all higher command echelons. There is increasing recognition that PSYOP need not be limited to the hardware-supported missions of the tactical battlefield but can have important applications at the operational and theater levels, particularly in low-intensity conflict situations. But there is as yet little apparent consensus on the role of PSYOP at the strategic level or in peacetime.

The tendency to think of PSYOP in terms of direct verbal communication is a strong one, and it reflects the nature of tactical PSYOP as historically practiced by the United States. However, this is a tendency that must be resisted if the full potential of nontactical PSYOP is to be realized and if the services are to embrace the full range of PSYOP activities as legitimate and proper military missions. The uniformed military generally acknowledge that the overriding purpose of US military forces is not to fight wars but to deter them. But deterrence is a psychological phenomenon, not a simple reflection of the quantity and quality of military forces; and there is every reason to suppose that foreign perceptions of US military power can be shaped in various ways to strengthen its deterrent effect.[22]

Even if one accepts that it would be difficult to shape Russian and Chinese perceptions of US power, a strong case can be made for the potentially high payoffs of efforts to shape perceptions in third world countries. Within the third world, decision making is apt to be undisciplined, inadequately supplied with intelligence, lacking in orderly staff procedures, strongly influenced by the passing impressions and phobias of a small leadership element, and subject to sudden internal political challenge.

Much could be accomplished simply through deliberate exploitation of normal US military activities such as exercises, deployments, air and naval displays, and technology demonstrations. At a higher level of activity, with the movement of military forces dedicated specifically to a psychological-political mission, PSYOP measures shade into traditional coercive diplomacy. With activities such as naval port visits and presence, missions have generally not been understood as belonging within the PSYOP framework and do not appear to have been approached in a systematic or highly coordinated manner.

A characteristic weakness of America's approach to the use of force has been the tendency to draw sharp distinctions between wartime and peacetime. One result is the penalty exacted (in terms of organization, planning, and general readiness) in any transition from peace to war. Psychological-political warfare could compensate for temporary inadequacies in deployed US forces in severe crises or in the initial stages of war. In fact, this could be an important function for the United States. In general, a compelling case can be made for reviewing and enhancing the psychological-political component of US war planning and national-level crisis management operations.

The entire area of strategic war planning is of critical importance in this context, since it is the point at which the military, diplomatic, and psychological-political components of national strategy most closely converge. Efforts to enhance and better integrate strategic planning at the national level should focus on the difficult substantive and procedural issues involved in war planning. More generally recognized is the need for integrated interagency planning in crisis situations; but here as well the potential of psychological-political warfare seems not to have been fully realized.

The foregoing discussion highlights the part of this field that is the most neglected and at the same time the most susceptible to immediate improvement. In low-intensity conflict theaters such as Central America, there is scope for application of the full range of US psychological-political capabilities. The strategic importance of peacetime political warfare and international communications—with respect to the third world as well as the major powers—can hardly be overestimated. It seems clear that the best hope for diminution of political-military threats from major powers lies in the relentless exposure of their populations to information and ideas from the West. The opportunities for short-term gains should not be allowed to distract us from this fundamental strategic imperative.

Notes

1. See, for example, Paul Linebarger, *Psychological Warfare* (Washington, D.C.: Infantry Journal Press, 1948); Leonard Doob, *Public Opinion and Propaganda* (New York: Henry Holt and Co., 1948); Daniel Lerner, ed., *Propaganda in War and Crises* (New York: Stewart, 1950); William Daugherty and Morris Janowitz, eds., *A Psychological Warfare Casebook* (Baltimore: Johns Hopkins University Press, 1958); and W. Phillips Davison, *International Political Communications* (New York: Praeger, 1965).

2. On the early history of military psychological operations, see Alfred H. Paddock, Jr., *U.S. Army Special Warfare: Its Origins* (Washington, D.C.: National Defense University Press [NDUP], 1982).

3. The originally top secret report of the president's Committee on International Information Activities, headed by William H. Jackson, is now available in Department of State (DOS), *Foreign Relations of the United States, 1952–1954*, vol. 2, pt. 2 (Washington, D.C.: Government Printing Office [GPO], 1984), 1795–1867. Hereafter referred to as *Foreign Relations*.

4. Ibid., 1836–60. According to the Jackson committee report,

> the national information program has suffered from the lack of effective central direction. In spite of the establishment of the Psychological Strategy Board, coordination has been lacking and the various agencies concerned have largely gone their separate ways. Opportunities have been missed to take the offensive in global propaganda campaigns. . . . The headquarters staffs of all agencies engaged in information work should concentrate more on the conception, planning, and coordination of global campaigns and less on detailed control and execution of day-to-day operations (1840).

For pertinent remarks on the confusion regarding the mission of US information agencies, see pages 1836–38; also of interest is this remark concerning the military role:

The contribution of the armed forces to political warfare has been limited by the lack of definition of the military role by higher authority, and by an inadequate understanding on the part of military authorities that they and their commands are full participants in the political aspects of the present struggle and must conduct themselves accordingly. Military commanders and planners tend to regard the allocation of military resources to current political operations as an unauthorized diversion from tasks for which the armed forces are explicitly responsible, (1860).

Eisenhower abolished the Psychological Strategy Board on the committee's recommendation, apparently on the grounds that psychological-political warfare was too closely bound up with US policy or strategy generally for such an arrangement to work effectively.

5. Consider, for example, John Franklin Campbell, *The Foreign Affairs Fudge Factory* (New York: Basic Books, 1971), 147–77.

6. A thoughtful discussion of the functioning of public diplomacy under the Reagan administration is provided by Gifford D. Malone, "Functioning of Diplomatic Organs," in Richard F. Staar, ed., *Public Diplomacy: USA versus USSR* (Stanford, Calif.: Hoover Institution Press, 1986), 124–41. See also Carnes Lord, "In Defense of Public Diplomacy," *Commentary*, April 1984, 42–50.

7. The argument for a more political approach to multilateral diplomacy is cogently summarized by Richard S. Williamson, "U.S. Multilateral Diplomacy at the United Nations," *Washington Quarterly*, Summer 1986, 5–18. On the Soviet approach to political action, see, for example, James Atkinson, *The Politics of Struggle: The Communist Front and Political Warfare* (Chicago: Henry Regnery, 1966); and Roy Godson, *Labor in Soviet Global Strategy* (Washington, D.C.: National Strategy Information Center, 1984). On the promotion of democracy as an objective of US policy, see William A. Douglas, *Developing Democracy* (Washington, D.C.: Heldref Publications, 1972).

8. See, for example, James Cable, *Gunboat Diplomacy: Political Applications of Limited Naval Force* (New York: Praeger, 1971); and Barry M. Blechman and Stephen S. Kaplan, *Force Without War: U.S. Armed Forces as a Political Instrument* (Washington, D.C.: Brookings Institution, 1978).

9. See Richard H. Shultz and Roy Godson, *Dezinformatsia: Active Measures in Soviet Strategy* (Washington, D.C.: Pergamon-Brassey's, 1984). Soviet active measures *(activnyye meropriatia)* encompass what is here called political action as well as covert political warfare. For a general discussion, see Angelo Codevilla, "Covert Action and Foreign Policy, II," in Roy Godson, ed., *Intelligence Requirements for the 1980s: Covert Action* (Washington, D.C.: National Strategy Information Center, 1981), 79–104.

10. A collection of (generally older) materials on this subject is available in Ronald D. McLaurin, ed., *Military Propaganda: Psychological Warfare and Operations* (New York: Praeger, 1982).

11. A discussion of the history of US international broadcasting is available in James L. Tyson, *U.S. International Broadcasting and National Security* (New York: National Strategy Information Center, 1983). For an overview of current US activities in this area, see United States Advisory Commission on Public Diplomacy, *1986 Report* (Washington, D.C.: GPO, 1986). Examples of public diplomacy materials produced by US government agencies are DOD, *Soviet Military Power 1986* (Washington, D.C.: GPO, 1986) and DOS and DOD, *The Challenge to Democracy in Central America* (Washington, D.C.: GPO, 1986).

12. The role of information as an instrument of national strategy comparable to the diplomatic, economic, and military instruments was emphasized in an address of then National Security Advisor William P. Clark to the Center for Strategic and International Studies, Georgetown University, 21 May 1982. Administration studies of international information policy resulted in a presidential directive on this subject (NSDD 130) in March 1983. See "Tune Up for Term Two," *Chronicle of International Communication*, July–August 1984, 1.

13. For some critical reviews by recent participants, see the remarks of Alfred H. Paddock, Jr., and Gifford D. Malone in Staar, 297–99.

14. Sun Tzu, *The Art of War*, trans. Samuel B. Griffith (Oxford: Oxford University Press, 1971), 77–78.

15. Gen Dwight D. Eisenhower told a meeting of American newspaper editors in 1944: "I have always considered as quasi-staff officers, correspondents accredited to my headquarters." That reporters acquiesced in such treatment is scandalous for a contemporary journalist; see the discussion in Philip Knightly, *The First Casualty* (New York: Harcourt Brace Jovanovich, 1975), 315–17ff.

16. The gross abuses involved in media coverage of the Vietnam War have been well documented. See Peter Braestrup, *Big Story* (New Haven, Conn.: Yale University Press, 1978); Robert Elegant, *How to Lose a War: The Press and Viet Nam* (Washington, D.C.: Ethics and Public Policy Center, 1982; originally published in *Encounter*, August 1981).

17. An excellent brief analysis may be found in Mark Crispin Miller, "How TV Covers War," *The New Republic*, 29 November 1982, 26–33. Also of considerable interest is John Weisman, "Why TV Is Missing the Picture in Central America," *TV Guide*, 15 September 1984, 2–14.

18. The controversy arising from the exclusion of the press from the US action against Grenada in October 1982 led the Joint Chiefs of Staff to initiate a review of policy regarding media participation in contingency operations, in the form of a special commission of military and media representatives headed by Gen Winant Sidle. Although this commission reached informal agreement on the principle of press participation in all military operations to the maximum degree consistent with operational security, the DOD has since indicated some reservations on this score, and there is little evidence that any significant movement has occurred on the media side. As one retired admiral is reported to have said, "Operational

security is not the problem; the problem is that when you write about us, you make us look bad." See Fred Halloran, "The Pentagon and the Press: The War Goes On," *New York Times*, 29 January 1986; and "Material and the Media," *Essays on Strategy and Diplomacy* 2 (Claremont, Calif.: Keck Center for International Strategic Studies, 1984).

19. Pertinent observations on the role of the State Department in the current administration's public diplomacy efforts are offered by Gifford Malone in Staar, 132–37.

20. Consider the complaints of Fred W. Walker, "PSYOP Is a Nasty Term—Too Bad," *Air University Review* 28 (September–October 1977), 71–76, as well as Edward N. Luttwak, "Notes on Low-Intensity Warfare," in William A. Buckingham, ed., *Defense Planning for the 1990s* (Washington, D.C.: NDUP, 1984), 197–209.

21. See the discussion in Alfred H. Paddock, Jr., "Psychological Operations, Special Operations, and US Strategy," in Frank R. Barnett, B. Hugh Tovar, and Richard H. Shultz, eds., *Special Operations in U.S. Strategy* (Washington, D.C.: NDUP, 1984), 231–51.

22. Of interest in this connection is Robert Jervis, Richard Ned Lebow, and Janis Gross Stein, *Psychology and Deterrence* (Baltimore: Johns Hopkins University Press, 1985).

When Are Political Objectives Clearly Defined?

A Historical Perspective

Michael A. Morris

National policy objectives should be clearly defined. Whether the goals are strategic or tactical, they must be delineated and articulated to those responsible for implementing them.

Since the difference between clarity and vagueness in political objectives is only a matter of degree, incomplete conceptions of vague political objectives obviously leads to incomplete conceptions of clear political objectives. Nevertheless, a rough distinction can be made between vague and clear political objectives after the nature of their relative difference is understood. Most observers acknowledge that political objectives should determine the nature of military objectives and strategy— which cannot be precise unless political objectives are clearly defined.

Clarity in political objectives is the responsibility of political leaders rather than military authorities; therefore, the military services cannot be expected to impose limits on their operations if the political objectives are vague.[1] In turn, if military power is to be used efficiently and without risk of major war, political objectives must not underestimate or overestimate the military capabilities at their disposal.

Correct Generalizations

The attempts of many authors to define clear political objectives are less perceptive than their analyses of the evils that result from objectives that are vague. Most envisage clear political objectives derived from an overall objective that reflects our national purpose. They point out that limited objectives determine the nature and extent of US involvement in limited wars or interventions. Some authors add that clear political objectives also need to be flexible so they can be adjusted to meet changing conditions.

These statements about clear political objectives are correct generalizations. Clear political objectives should be limited and should not be achieved at the price of inflexibility. Changing circumstances will always make the task of formulating clear political objectives difficult. However, attempts to go beyond generalities belie incomplete conceptualizations of vague political objectives.

A Paradox

Political objectives evince much superficial continuity and, hence, an aura of decisiveness, although they may actually encompass and conceal considerable drift and indecision. Thus, while flexibility is necessary for adapting to changing circumstances, flexibility can also become largely synonymous with drift and indecision.

Nevertheless, political objectives still may remain constant in meaning over a period of time although different strategies may be required to implement them. The difficulty is in distinguishing between political objectives that only have superficial continuity of meaning and those that really are continuous in meaning.

This paradoxical nature of vague political objectives suggests a definition of clear political objectives. A political objective is clear when it continues to represent concern for the preservation of the same specific vital interest or for the same kind of postwar peace. If the objectives are clear in this sense, they can provide adequate political direction for the choice of military objectives and strategy.

Drift occurs when political objectives are not continuous in meaning or when they are so vague as to permit several divergent strategies to appear to be adequate ways of attaining objectives. In either case, drift rather than political objectives largely determines the nature of strategy.

However, this definition applies only to those specific political objectives that guide military strategy in a specific limited war or intervention. General foreign policy objectives are, perforce, vague. For example, support of the United

Nations and interest in a world free of aggression are general foreign policy objectives that attract popular support through their vagueness.

This suggested definition of clear political objectives needs to be elaborated further by examining its relation to the concept of vital interests. Since clear political objectives must represent concern for the preservation of the same specific vital interest, it is then necessary that vital interests also be clear and specific.

Propaganda Advantages

Some propaganda advantages can be gained by grandiloquent political objectives such as the Truman doctrine's aspiration to support free peoples who are resisting attempted subjugation by armed minorities or outside pressures. However, even at that early date in the cold war, the rhetorical exaggerations involved touched off a debate charging that the United States was overcommitting itself.

The US strategy of support to Greece and Turkey against aggression refuted these criticisms; it was a clear and limited commitment. However, this clarity of strategy was due to the US policymakers' clarity of purpose and their clear grasp of US vital interests rather than to the clarity of publicly stated political objectives.

Occasionally, vital interests may be precisely formulated while political objectives are vague; for example, the Truman doctrine. Nevertheless, it is extremely difficult for political objectives to be clear when vital interests are abstract and vague.

The stark clarity of a direct Soviet confrontation with the West in the late 1940s made the determination of US vital interests relatively simple. The calculation of what were these vital interests became more complex and ambiguous as the communist threat became more diverse and subtle in the 1950s.

Abstract Vital Interests

As the paradox of abstract vital interests began to loom larger in US policy in the 1950s and 1960s, the need for clarity of political objectives continued and, perhaps, even increased as the nature of the threat changed and became more ambiguous. However, the precedent of vague political objectives set by the Truman doctrine's rhetoric continued to be the observed norm.

One of Robert E. Osgood's main themes in a 1968 paper illustrated the consequences of a heritage of vague political objectives and of increasingly vague and abstract vital interests. While the US postwar pursuit of policy objectives has shown much continuity, the US concept of its vital interests has been constantly expanding.

In Vietnam, where communist aggression was considered a threat to US security, the general requirements of containment led to an increasing US response with no question as to the precise relevance of the Vietnam War to US security. Osgood argued that the growth of multipolarity and the increasing complexity of the communist threat necessitated more precision in determining the extent to which our vital interests are involved in communist aggression.[2]

The paradox of abstract vital interests, then, reinforces the paradox of vague political objectives and vice versa. Together, they reinforce the indecision embedded in each vague paradox. This tends to result in policy drift and in inadequately considered, expanding commitments.

This vicious circle of vagueness and drift can be broken by efforts to relate vital interests more specifically to contemporary conditions and conflicts, and can be weakened by increasing the clarity of political objectives; that is, clarification of vital interests leads to clarification of political objectives and vice versa. More precise vital interests make it harder for political objectives to equivocate in claiming they are preserving the same specific vital interest.

On the other hand, greater clarity of political objectives would increase pressures on vital interests to be more precise. Greater clarity of political objectives would indicate the respective advantages and costs of implementing different strategies and combinations of political objectives, which would in turn increase the chances that vague concepts of vital interests would be exposed. Such exposure would surely result in greater pressure for more precise definitions of vital interests.

Guidelines

The peculiar characteristics of vague political objectives suggest several guidelines for distinguishing clear political objectives from vague ones. Clarity of political objectives is being achieved when the following questions can be answered in the affirmative.

Is the nature of the paradox of vague political objectives understood and consciously guarded against in defining political objectives?

Do political objectives help undermine the paradox of abstract vital interests? By indicating the respective advantages and costs of implementing different strategies, do political objectives expose vague conceptions of vital interests that unwisely require prohibitive costs?

In choosing our military strategy, have we carefully considered the costs and benefits of attaining objectives that primarily demand military action versus the costs and benefits of attaining objectives that are more closely related to nation-building and internal security?

Are drift and indecision being reduced in our political objectives, our conception of our vital interests, and our choice of military strategy?

Notes

1. Robert J. Bower, "Military Objectives in the Nuclear Age," *Military Review*, May 1966, 91–97.

2. Dr Robert E. Osgood, "Perspective on American Commitments," paper presented at the Sixty-fourth Annual Meeting of the American Political Science Association, Washington, D.C., 5 September 1968, 11, 17–19.

Some Thoughts on Psychological Operations

William F. Johnston

> *Governments of countries threatened with insurgency should regard PSYOP, particularly face-to-face communications, as a first line of internal defense.*

The intent of this article is not to proclaim PSYOP a panacea for "wars of liberation."[1] The intent is to focus attention on psychological operations as a vital instrument in Ho/Mao-type wars of national liberation, especially the early stages where revolutions are hatched from grievances of the masses by communist incubation and kept at the right emotional temperature by thousands of native agitators.[2] Without this skillful and massive agitprop and organizational effort, which has been characterized as "half the revolutionary task," there could be no successful liberation wars.

Auxiliary or Primary?

Psychological operations should have more recognition as important instruments of low-intensity conflict.[3] They have the capability to compete with the communists' conflict doctrine, which requires the integration of political, economic, psychological, and military factors of power. Aggressive nations can successfully use PSYOP on a case-by-case basis in subversive insurgency-threatened countries. PSYOP use, however, must be carefully calculated in policy and operations to reach the grassroots level. Whoever gets to the people first, with ideas that stimulate self-interest, gains a decisive lead . . . and that is precisely what happened in Vietnam.[4]

Use of Media Discounted

Looking at the Vietnam War in retrospect, we see that the communists used all available media, from instigation through propaganda and guerrilla warfare to conventional military operations. Such sophisticated mass media as printed materials, leaflets, posters, pamphlets, and radio broadcasts were more obvious than the workhorse of the revolution: the agitprop cadre who operated on a face-to-face basis in the rural villages and hamlets. The US discounted the sophisticated media in the early days because it did not seem to play an important role.[5] What no one understood at the time was that radio was being used to communicate the Communist party line to the agitprop cadres in the remote areas of South Vietnam. They, in turn, were translating the party line into action among the masses. Based on opportunities at the local level, the actions generally consisted of study sessions (learning how to read and write Vietcong propaganda), demonstrations, parades, rallies, and any other activities that would get the people involved—ostensibly in furtherance of the peoples' interests, but with the end objective of strengthening the Communist party and weakening the opposition. The US did not appreciate the magnitude of the agitprop cadres—their numbers, their effectiveness, or the extent to which they were able to get the people engaged in furthering communist propaganda.

Douglas Pike, a noted expert, claims that the various social organizations created by the Vietcong in South Vietnam and used by the agitprop cadres were developed as "self-contained, self-supporting channels of communication." In his view, these organizations were the "secret weapon" and the "heart and power of the National Liberation Front (NLF)."[6]

Pike also revealed that the Vietcong "spent enormous amounts of time, energy, manpower, and money . . . explaining itself to itself, to the other side, and to the world at large." He said they were "obsessed" with doing this. Pike found that the more he studied the Vietcong, the more "it became evident that everything the NLF did was an act of communication."[7] In other words, everything the Vietcong did was calculated to gain the optimum psychological effect.

Policymaker and Propagandist

In the communist world, psychological operations/warfare experts are represented at the highest levels of government and within the Communist party. One of their roles is to help make policy. PSYOP experts in the United States and most of the free world, however, (except for Nationalist China) appear to have little status and not much influence; several reasons are offered below.

1. Psychological operations/warfare, including propaganda, has a distasteful image. It is thought to deal in lies and variations of the truth, and is looked upon in general as a dirty business with which few wish to be associated.

2. Psychological operations cannot be scientifically evaluated as to effectiveness. PSYOP activities that cannot produce fast, measurable results are hard to sell. There are exceptions, of course; for example, when tactical PSYOP can be used to encourage a beaten enemy to surrender or in an amnesty campaign where the defectors can be counted.

3. There are very few professionally qualified and experienced PSYOP experts who also have sufficient acquaintance with the people of a specific foreign country. A major reason is that little status, low prestige, few rewards, and no high-level promotions are based on professionalism in psychological operations. There is no psychological operations career program; yet, attaining full qualification as a psychological operations expert on a specific foreign country is a lifetime job, not unlike the type of professionalism required of a psychoanalyst.

4. Insurgency-threatened countries do not give enough attention or importance to psychological operations. They often lack the know-how to start an effective psychological operations program, and they tend to place the same priority on PSYOP that the United States does in its support of that country.[8] This priority is usually relatively low in proportion to the overall assistance provided by the US. While in the later stages of an insurgency the US may help with such sophisticated media as printed leaflets, radio, and TV, the US usually lacks expertise in the language and ways of the people. This expertise is necessary to provide advice on the content of the message to ensure that the message achieves the desired

effect. The main burden of face-to-face education must be on the shoulders of the indigenous government and people.

Another problem is that of convincing an insurgency-threatened government to establish an uncomplicated but credible amnesty program early in the insurgency. This program would provide, for insurgents who have changed their minds, a way out of the "victory or death" box that the guerrilla is usually in.

Chieu Hoi

The Chieu Hoi (Open Arms) Amnesty Program was one of South Vietnam's most successful programs. However, considerable time was required for the program to build up credibility and to achieve the desired results.[9] Initially, Vietcong who wanted to defect believed, as the Vietcong indoctrination claimed, that they might be shot.

After the low point at the end of 1964, the Chieu Hoi program showed a steady increase in the number of Vietcong returnees. In 1966 there were over 20,000 defectors, double the number of the preceding year. Total defections of Vietcong returning under this program numbered more than 75,000.[10] If we accept the ratio of 10 government soldiers needed for each insurgent guerrilla, this program saved the GVN and the US a troop strength of over 750,000 soldiers. From the dollars-saved angle, the total cost of the program, using a figure of $127 to bring in a Vietcong defector, was around $9.5 million. Since the cost to kill a Vietcong is estimated at $300,000, killing this number of soldiers would have cost $2.25 billion.

Indigenous Armed Forces

The United States can and should give a higher priority to encouraging insurgency-threatened governments to develop a professional psychological operations capability. This capability must include personnel, both American and indigenous, who can conduct face-to-face psychological operations. Indigenous PSYOP people must be able to overwhelm communist-trained agitators in both quality and quantity of ideas related to peoples' desires and fears if they are to get the people motivated and committed to the government.

Indigenous military forces, paramilitary forces, or police forces would be good organizations to undertake increased PSYOP responsibilities in the rural areas. One reason is that these forces represent the best organized and most cohesive institutions in many developing countries, particularly in the rural areas. Also, better results will be achieved in the remote rural areas if the peasants know that an iron fist is underneath the velvet glove of the communists' friendly persuasion. (Changing the ways of peasants is a real challenge.)

Every member of those military and police forces that are in contact with rural people should be trained to talk to the people—and doing so should be required as part of their duties. This would include interpreting news and participating in educational activities. They must also be prepared to discuss persuasively what the government is doing for the local people.

The main focus should be on developing specialized PSYOP units in the rural areas. These units should consist of local natives. Very important, also, is the establishment of a two-way communication system to ensure that popular grievances and good ideas get to the PSYOP planners and decision makers at the top level—and that governmental PSYOP policy guidance gets down to the lowest PSYOP level. This policy guidance must be centralized and controlled at the top, but implementation of the policy must be decentralized at the grass roots.

Lower-level PSYOP people should spend about one-half of their time in investigating and preparing so as to tailor their talks and educational activities to fit the local needs. Once the real needs of the rural people are ascertained and transmitted to the top echelons, there is more reason to hope that the government will help the people help themselves. This would go a long way toward preventing the communist agitators from generating revolutions out of grievances, hatreds, and social injustices.

Conclusions

In conclusion, insurgency-threatened governments can be expected to heed US advice if it is in the form of lessons learned at great cost. Indigenous governments must put a high priority on psychological operations, particularly on face-to-face education and on countering communist agitators. These governments must recognize that a face-to-face psychological operations capability must be developed from the people, by the people, and for the people of each particular language group, tribe, clan, or area of the country—and this takes time. Developing a successful capability requires careful selection and training of candidates. This is not something that can be "made in the USA" and exported.

The US military can and should develop mature, fully qualified, professional PSYOP experts who are experienced in the language and the thinking of a foreign people. Such experts could help governments with PSYOP organization and management, as well as providing support with our more sophisticated mass media. Psychological operations planners, both US and indigenous, *must* be raised to the first-team level. This is crucial for success! PSYOP planners must have ready access to top authorities. Governments must translate top-level interest and support into meaningful terms for people at the grassroots level. Feedback from the people to the top level is equally important. This interest from government is essential to winning and maintaining the support and loyalty of the people. Such a psychological operations program would be one of the cheapest and best security investments the US could make in the developing world.

Notes

1. This article consists of excerpts from "Neglected Deterrent: Psychological Operations in 'Liberation Wars'," *Transition*, nos. 12 and 13 (January 1968): 58–65. Reprinted with the permission of the Foreign Affairs Executive Seminar, Foreign Service Institute, and its author. The communist psywar instrument was simply a means to an end—communist domination over everything that was not already under communist control.

2. Late in the China conflict, Gen George C. Marshall realized that the battle for the mind waged with ideas and propagated by mass communications media could be decisive in countering the liberation war in China. In

1947, upon return from his unsuccessful mission to China, Marshall said, "China might have been saved by the massive use of radio and motion pictures, on a scale hitherto unheard of." William Benton, "How Strong Is Russia? And How Weak?" *New York Times Magazine*, 10 June 1956, 70.

3. Robert Holt and Robert Van de Velde, *Strategic Psychological Operations* (Chicago: University of Chicago Press, 1960), 64. These authors state that the US has never fully understood the nature of the psychological instrument.

4. Stefan Possony in "Viet Cong Propaganda War," *Los Angeles Times*, 11 July 1967, gives a good analysis of international propaganda based on captured documents. He concludes that although Americans are not conscious of it, the US was the target of a well-orchestrated and skillful Vietcong propaganda offensive after 1961.

5. Radio Hanoi and Vietcong clandestine transmitters were important for getting the Communist party line and other instructions to the leadership and agitprop cadres in remote areas—information which otherwise would have been delayed for weeks. Broadcasts were made in all five major languages used in Vietnam. Beginning in 1960, there was an increase in the use of all types of media in an accelerated propaganda offensive against the government of Vietnam.

6. Douglas Pike, *Viet Cong: The Organization and Technique of the National Liberation Front of South Vietnam* (Cambridge, Mass.: MIT Press, 1967), 124.

7. Ibid., ix.

8. See Slavko Bjelajac, "A Design for Psychological Operations in Vietnam," *Orbis*, Spring 1966. Bjelajac concludes that "psychological operations are indispensable" and in a Vietnam-type war "should be accorded a priority at least equal to any other weapon or technique in the Vietnamese protracted conflict."

9. The task of defecting a Vietcong in the Chieu Hoi Program was primarily a psychological operations task. However, Aid for International Development (AID) Vietnam played an important role, since AID furnished the resources for most of the food, housing, and allowances given to the Chieu Hoi.

10. Bob Considine, "Pacification Cadres," *Philadelphia Inquirer*, 19 September 1967.

The Role of Public Opinion

Lloyd A. Free

Public opinion cannot be slavishly followed, but psychological data should be collected and analyzed so that government can take this factor into account in planning.

The following excerpts are from an exchange of letters that took place at about the time of the Vietnam moratorium in the fall of 1969. One writer was a Georgetown University sophomore with the unknown name of Randy Dicks, the other a president of the United States with the well-known name of Richard Milhous Nixon.

Randy wrote the president:

I think that your statement at your recent press conference that "under no circumstances" will you be affected by the impending antiwar protests in connection with the Vietnam moratorium is ill-considered, to say the least. It has been my impression that it is not unwise for the President of the United States to take note of the will of the people. After all, these people elected you. You are their President.

The president replied, in part:

There is a clear distinction between public opinion and public demonstrations. To listen to public opinion is one thing; to be swayed by public demonstration is another. . . . Whatever the issue, to allow Government policy to be made in the streets would destroy the democratic process. It would give the decision not to the majority and not to those with the strongest arguments, but to those with the loudest voices.

Introduction

What is the role—the actual and proper role—of public opinion in international security affairs? And how, if public opinion does and should count, is majority opinion to be determined?

The assumption that public opinion, both at home and abroad, is somehow important is borne out by the efforts of political leaders to woo it and by the practices of governments to influence it. All major governments in the world today, and many of the minor ones, spend varying amounts of time, money, and attention on attempts to influence the opinions of their own citizens and the citizens of other countries.

Yet, even in our recent history, when a sense of the importance of public opinion has become more self-conscious than it used to be, there have been unbelievers. Harry Truman, whom I nevertheless admire as one of our great presidents, was one of them. When Mr Truman faced a problem, he would find a principle involved, often a moral principle, and then make his decision accordingly. And he would stick to that decision come hell or high water. With this approach, public opinion and opinion polls are irrelevant; you simply do what you think is right.

The late, great John Foster Dulles adopted this same approach. He once said in my presence:

> If I so much as took into account what people are thinking or feeling abroad, I would be derelict in my duty as Secretary of State.

Another of our great secretaries of state, Dean Acheson, claimed, disapprovingly, that Americans have a "Narcissus psychosis." "An American," he wrote, "is apt to stare like Narcissus at his image in the pool of what he believes to be world opinion." After making the point that the only honest answer people generally could give to questions about the specifics of foreign policy would be a "don't know," he made this observation:

> World opinion simply does not exist on matters that concern us. Not because people do not know the facts—facts are not necessary to form opinions—but because they do not know the issues exist.

Thus, we are faced with some very basic questions: Does such a thing as world opinion exist? Do people in the United States and other parts of the world really have meaningful opinions of any significant scope in regard to international issues? If so, are these opinions of importance to foreign relations and international security affairs? Your own instinctive answer to these questions may be an unqualified "yes" or an unqualified "no," but after years of experience in the public opinion field, my own answer is very equivocal—namely, it all depends.

Definitions and Assumptions

To start with, I must define some of the terms we will be discussing. An opinion, in my terminology, is simply an expressed attitude—an attitude that is communicated. An attitude, on the other hand, is really more of a perception—a way of looking at a given subject.

In the course of our lives, we build up all sorts of assumptions based on our experiences and these assumptions influence what we perceive as the world in which we live. In other words, we participate in creating our own realities and our attitudes spring from these "realities." They are the result of an interplay of our assumptions, as shaped and modified by experience. In a very real sense, if an individual has no assumptions concerning a given subject, or that are capable of being related to that subject, he can have no attitudes—and hence no opinions. And any opinions expressed by an individual will be meaningful only if he in some way relates the subject to his own purpose, no matter how narrow or broad it may be. The range of his sense of purpose is delimited by his "reality world."

Every individual has blind spots of greater or lesser scope—that is, subject matter areas about which he has no assumptions and, hence, no attitudes. It is difficult for many people to realize what a large proportion of the people of the United States, not to mention the underdeveloped areas of the world, have no assumptions, attitudes, or information about international affairs.

Let me cite from a study I did in this country a few years ago. One-quarter of the American public had never heard or read of NATO, only 58 percent knew that the United States is a member of NATO, and only 38 percent knew that the Soviet Union was not a member—facts which go to the very nature and fundamental purpose of our most important alliance! One-quarter of the adults in this country did not even know that the government of mainland China is communistic!

In short, at least two-fifths of even the American people are far too ignorant about international affairs to play intelligent roles as citizens of a nation that is the world's leader—and only about one-fourth are really adequately informed. The situation in most other countries of the world is far worse, particularly in underdeveloped areas. For pollsters to ask these uninformed people about specifics of foreign policy is obviously an exercise in futility. Looked at in this perspective, one can begin to see the validity of Dean Acheson's views and to question the common assumption that, if enough people at home and abroad are persuaded to adopt a given opinion, then the policy of their government will be affected—at least in democracies.

Opinion Leaders and the General Public

Before we write off the importance of public opinion in international security affairs, however, let us introduce some other aspects of the problem. First, we must recognize that when it comes to the "nitty-gritty" day-to-day decisions on specifics, public opinion usually has no effect. Either the public has no opinions on the matter or people do not know that decisions are being made. On a broad range of matters that are publicly known, there is usually an educated elite who do have meaningful opinions, in varying degrees of intensity, about international security matters.

This elite may be of greater or lesser size, depending upon which country is involved and which issue is in question. However, the fact that it may be small does not derogate its power. We can meaningfully define world opinion abroad or significant opinion at home in terms of the publics that count in the particular situation, whether limited or mass.

Beyond this, however, elements of the public can and often do get into the act, not only in the United States, but also in the underdeveloped areas. The people may lack meaningful opinions on a wide range of specifics about international matters, but their broader assumptions may come into play at certain times and places to make a given international issue a matter of public concern. Often this applies only to a minority of the greater public; frequently the people are whipped up and organized for ulterior ends, whether by the communists or by local leaders—but react they do!

Publication

The action may be as peaceable as signing a petition or writing a letter to the local newspaper (both of which are apt to be relatively ineffective), or to writing a personal letter to the president (as Randy Dicks did), or writing to one's congressman. (The latter action may actually be of some influence; the attention paid by members of Congress to their mail is out of all proportion to its significance as a barometer of public opinion.)

But, increasingly, more extreme manifestations of public action—demonstrations, picketing, and rioting—reflect attitudes that are strongly held by at least some segments of the public. These forms of public action have become a phenomenon of worldwide scope. For example, rioting in Japan and Korea made it exceedingly difficult for the two governments to normalize their relations. Demonstrations in Panama were unquestionably instrumental in causing the US—after a decent interval, of course—to agree to revise the Panama Canal Treaty. Demonstrations and potential riots in the Middle East have made it difficult for the Arab governments to follow a policy of moderation in the Arab-Israeli conflict. Anticommunist violence in Indonesia strengthened the hand of the army against the Communist party in a struggle that has had profound international implications. Anti-Vietnam protests in the United States unquestionably affected the calculations of the North Vietnamese, not to mention those of our American leaders.

But the greater public also gets into the act in a more regular and generally more peaceable way in the form of periodic elections, not only in the democracies but also in some of the semidemocracies (if not the "guided democracies"). In such elections, international matters can and often do enter as central issues of the campaign.

But then, one does not have to search very far back in US history for other examples: Woodrow Wilson's campaign theme, "He kept us out of war," helped him win the election in 1916; and Gen Dwight D. Eisenhower's promise to go to Korea increased his landslide victory in 1952. In fact, it is almost axiomatic that whenever war or peace seems to be at issue, the public in almost every country will exhibit deep concern in ways that have political meanings.

Global Public Opinion

More broadly, a close study of the matter has convinced me that there are widely shared attitudes on international matters, amounting in many instances to consensuses which governments simply *must* take into account. Sometimes these consensuses are global. With apologies to Dean Acheson, there are, on occasion, worldwide or virtually worldwide reactions on matters that concern us.

One occasion was the Suez affair in 1956. I have little doubt that the United Nations well-nigh universal condemnation of the Israeli-British-French invasion of Egypt was supported by what can only be called a consensus of world opinion—a consensus shared even by many people in both the United Kingdom and France.

A similar consensus of condemnation seems to have existed over the Soviet Union's occupation of Czechoslovakia—a consensus apparently shared even by a good many people who were communists themselves.

Another is the worldwide impact of Russia's launching the first two sputniks in 1957, followed by its subsequent achievements in space. These developments led to reevaluations of the relative standings of the two superpowers, extending not only through official circles and elites but to general publics as well. In fact, data show that people throughout the world ranked the Soviet Union and the United States just about equally in terms of power and importance in those days, an enormous contrast from the days before 1957.

This perceived equality contributed, along with other developments, to the idea that a stalemate existed—a notion that affected the foreign policies of most of the world's nations.

Regional and National Attitudes

Short of these global consensuses, some basic attitudes are so widely held in certain regions or areas that they must be taken into account, both by the governments which rule there and by others dealing with them. The phobia in Latin America against American intervention is one example. Similarly, in almost all of Africa and Asia, basic attitude patterns opposed to imperialism and neocolonialism are deeply rooted. Other examples are the anti-Israeli "set" of the Arab world and, fortunately for us, the anti-Chinese bias in much of Southeast Asia.

In addition, there are many situations where there is a meaningful consensus of public opinion in particular countries. One example is the almost universal aspiration that the West Germans held for Germany's reunification. Other examples are the fear and hatred of Germans that were held by most Russians and Poles, and the Japanese public's opposition to full-scale rearmament.

The American people, too, have certain fixed ideas. One is opposition to foreign aid: Six out of 10 Americans favor either reducing economic aid to foreign countries or ending it altogether. Another is the very high degree of concern to keep our military defense strong at the same time that more than one-half of the public thinks we are spending too much on defense. In a related vein, almost six out of 10 Americans think the US "should take all necessary steps to prevent the spread of communism to any other parts of the free world, no matter where." Anticommunism is, in fact, clearly one of the strongest factors in Americans' ideas about US international security policies.

Policies and Opinions

In any particular country at any given time, there are programs and policies for which no government or leader can engender public endorsement. In other words, the climate of opinion imposes limits—sometimes very broad, sometimes very narrow—on each government's area of maneuver. In the extreme, certain things are virtually taboo; in other cases, they are merely impolitic; in still others, particularly where public opinion is either in agreement, nonexistent, divided, or lacking in intensity, anything is acceptable. Although policymakers sometimes appear blind to the fact, the achievement of many—if not most—of the international security objectives adopted by the United States presupposes certain perceptions, attitudes, beliefs, and behaviors on the part of various persons in this and other countries.

Usually, on day-to-day matters, the opinions of key members of Congress and/or officials of foreign governments are important for the accomplishment of particular US objectives. On more important matters, the list of people who have influential opinions may include members of elite groups that have power or influence in our own or other societies. Often, the educated elements of the general public may also have a bearing on the success or failure of US policies. Frequently, the opinions of peoples as groups are important, either because popular support or cooperation is necessary or because the people show their concern by way of protests, demonstrations, or elections.

Often, psychological factors are critical to the success of US policies. If these factors are absent, it is futile to adhere to a policy which presupposes them. Taking into account, for instance, the feelings of the majority of the Chinese people toward the Chiang Kai-shek regime in 1949, no amount of effort or determination on the part of the United States could have prevented a communist takeover of mainland China. As another example, consider that the US supported French rule in Algeria for years, despite the fact that the psychology of the Algerians—not to mention the French—made this goal impossible.

In another aspect of French international security affairs, it was public opinion, more than anything else, that forced the French military to fight the Indo-Chinese war with one hand tied behind its back; victory was impossible. The French people would not condone the necessary national effort, so the government found it impolitic to send draftees to fight in Indochina. The current malaise in the European community and the disarray in NATO are prime examples of the limits that psychological factors can place on national objectives.

Public Opinion, Political Pressure, and Prophecy

Finally, to prove what a valid prophet I can be on occasion—and, of course, I like to forget those instances where my judgment proved faulty—let me quote from a lecture I delivered when the trends of American opinion about the war in Vietnam were still pretty obscure.

> Our own Government will undoubtedly now have to face up to the fact that the American people are sick and tired of the war. It is my considered judgment as a so-called "expert" that we are in the early stages of an inexorable tide in favor of pulling out of Vietnam. There may be riptides from time to time which will temporarily obscure the direction of the current, but it is my belief that however you and I may feel about the matter, the movement down below will continue ever more strongly in favor of disengagement.[1]

Of course, as usual when I stick my neck out, I had some data to rely on. In the aftermath of the Tet offensive, studies conducted in this country showed that in mid-February 1968, immediately after the offensive, the majority of Americans remained hawks. In fact, hawkish sentiment increased in the immediate sense after the Tet offensive, favoring further escalation of the war. One-quarter of the American people advocated gradual escalation and no less than 28 percent opted for "an all-out crash effort in the hope of winning the war quickly, even at the risk of China or Russia entering the war."

By June of the same year—1968—the picture had changed materially. One-half of the public had shifted to the "dove" side, with 7 percent favoring a cutback in the American military effort and 42 percent wanting us to discontinue the struggle and start pulling out of Vietnam—this latter figure being almost double what it had been just four months earlier.

By June of 1970, Gallup had found that the proportion of people thinking that the US had "made a mistake in sending troops to fight in Vietnam" had risen from 25 percent (in March of 1966) to 56 percent. About one-half of the people now favored withdrawal, either immediately or at least by July 1971.

Policymakers and Public Opinion

There is a real question about the sensitivity of our policy-makers to psychological factors that have a bearing on the success or failure of our international security objectives. There is also a question about those policymakers' receptivity to the results of public opinion polls and other forms of policy-oriented psychological research. I can only speak on the basis of my own experience and that of my late associate, Hadley Cantril, plus years of personal observation of the government process. And from these points of view, I would say that the record is spotty at best. However, recent presidents have shown an awareness of the importance of psychological factors.

Oddly enough, the supreme example of sensitivity to public opinion—and of consummate ability to influence it—came in the earliest days of scientifically conducted polling. The man of the hour was Franklin D. Roosevelt. Tommy Corcoran once credited Roosevelt with having said, in effect, that he was the captain of the ship but that events and public opinion limited his power while providing instrumentalities for exerting power.

Particularly after the adverse reactions to his famous "quarantine" speech, Roosevelt was determined not to get too far out in front of public opinion in connection with the war in Europe—nor to stay any farther behind than he thought he had to. In this connection, he followed the polls with great interest—particularly charts of American public opinion specially prepared for him by Mr Cantril, who had conducted surveys throughout this period.

President Eisenhower

President Dwight D. ("Ike") Eisenhower was less consciously interested in domestic public opinion polls than Roosevelt had been. This was probably in part because Ike was so popular and his administration so relatively noncontroversial. But I know from my own experience that he was deeply interested in the opinions of people in other countries. While working with Nelson Rockefeller, who was a consultant to the president in

114

1955, I started a series of periodic reports on the psychological situation abroad.

The president read these reports carefully and followed them with great interest. (The common notion that he did not read documents is ridiculous; every single report I submitted to him through Rockefeller was read—and annotated—by the following morning. And this goes even for a 67-page document, single-spaced, which I know he read in full because he corrected a typographical error in his own hand on the next-to-last page!)

On more than one occasion, after John Foster Dulles had given one of his masterful briefings to the National Security Council (NSC), the president was heard to say, "But, Foster, you forget the human side," as he pulled out one of my reports and read from it. As a result of that, I was the second most hated man (by Dulles) in Washington, Rockefeller being the first!

By 1955, my reports to President Eisenhower had shown a sharp increase in skepticism abroad about America's peaceful intentions. This skepticism helped to create receptivity to an idea that Rockefeller had advanced and which had been cold-shouldered—namely, the "Open Skies" inspection proposal. When Eisenhower finally propounded Open Skies, it had as great a psychological impact as any one-shot propaganda move since World War II.

President Kennedy

Jerome Wisener says that John Fitzgerald Kennedy knew clearly that he could exercise his power only if he had the consensus of the people and the Congress behind him. As a result, Kennedy was a fervent believer in polling. He depended heavily on the findings of special surveys conducted for him by Lou Harris on domestic opinion, including international issues. He also followed closely the United States Information Agency (USIA) data on opinion abroad. On a number of occasions, he personally requested that certain surveys be made—especially in Latin America.

In the aftermath of the Bay of Pigs, USIA found that Fidel Castro was little known in Latin America and was generally viewed by the public with considerable allergy. Subsequently, the Kennedy administration adopted a relatively low-keyed approach to Castro and Castroism.

President Johnson

In the early days of his administration, Lyndon Baines Johnson regularly used Oliver Quayle's polls, including those on international issues. President Johnson kept a loose-leaf notebook, not only of the latest surveys taken in the United States by Quayle, Gallup, and Harris, but also of polls conducted abroad.

Following the American intervention in the Dominican Republic, I sent to the White House a report I had prepared in June of 1962, *Attitudes, Hopes and Fears of the Dominican People.* It showed that the Dominicans, as of then at least, were the most pro-American, anticommunist, anti-Castro people we had found in any part of the world. Not only did President Johnson read the report; the White House had it duplicated and distributed at the highest level and said that it had proved "very helpful." I have little doubt that this report was one of many factors that influenced the Johnson administration to shift from supporting the military junta to working toward a coalition solution.

Against such "successes" as these, however, must be set some glaring "failures," where research findings were ignored in framing US policies. The most cataclysmic of these had to do with the Bay of Pigs invasion. A year before, I had managed "by the skin of my teeth" to get a public opinion study done in Cuba. It showed that Castro was overwhelmingly popular with the Cuban people. There was a small opposition, but it was confined almost entirely to the city of Havana. Thus, whatever the expectations of those who planned the invasion, it came as no surprise to many that there was no popular uprising to assist the Bay of Pigs invaders.

The study had been made available to the government as well as to the public, and had actually been sent up to the White House. However, between the time the report was issued and the attempted invasion, there had been a change of administration—Kennedy had come into the White House. And our findings were not called to the attention of the new president or anyone on his staff when they were considering the question of invading Cuba.

Although called a "success" when submitted to President Johnson after the Dominican intervention in 1965, a Dominican study conducted after the fall of Trujillo found that the invasion had been a "failure" when the results were originally made available to the government in 1962. The report included the following flat statement:

> An extremely serious situation of popular discontent and frustration, fraught with a dangerous potential for upheaval, exists in the Dominican Republic. Never have we seen the danger signals so unmistakably clear.

Yet, despite this urgent warning, the US government devised well-merited but long-range solutions to Dominican problems and neglected short-term emergency programs which might have avoided later problems.

The neglect of such research findings demonstrates a fact which has become crystal clear to me from my years of experience within the State Department and my subsequent observations as a researcher and government consultant. A considerable proportion of our government, with particular reference to the State Department and the Foreign Service, tends to be insensitive to the importance of psychological factors in international affairs. Within our US government there is no systematic collection of such data and, despite the interest of the White House, no systematic marshalling of whatever material may be available. (It is not without significance that the special polling of American public opinion that was inaugurated by the State Department after World War II and was conducted under the very distinguished and able direction of Schuyler Foster, whom some of you probably know, was discontinued in the latter half of the 1950s.)

As to psychological factors abroad, more than one ambassador has echoed the words of one of their colleagues: "To hell with public opinion! I'm here to deal with the government, not with the public!" He said this, incidentally, while serving in a country which later went into an acute crisis because of public turbulence.

It is against this multifaceted background that I said the record of our policymakers is spotty when it comes to paying attention to "the psychological." Some policymakers (particularly at the highest level) have paid attention; some have not.

Conclusion

In closing, let's put this whole matter into broader perspective. No responsible critic maintains that our government should slavishly follow public opinion or that US foreign policy should be based exclusively, or even primarily, upon courting momentary popularity at home or abroad. Often, governments that deserve the adjective "responsible" will have to fly squarely in the face of domestic opinion while attempting to change it.

Bill Moyers once told a story that is pertinent in this respect: During a period of crisis, President Johnson and his advisors were meeting in the cabinet room to discuss alternative courses of action. At a particularly exasperating moment, when no option appeared likely to succeed, one of the men exclaimed wearily: "If we only knew what the people of this country really want us to do!" The president studied his melancholy advisor for 39 seconds, then answered: "If we knew what they wanted us to do, how could we be sure that we should do it?"

Any government worthy of its name has to do what it thinks necessary for the good of the country, but its course can be greatly eased if it has public opinion on its side. Decision making demands a knowledge of how the people are thinking and feeling—and why.

Even more often, governments must fly in the face of opinion abroad. This is especially true of the United States as it pursues its role of world leadership: if we aid India to rearm against the Chinese threat, we are bound to incur the wrath of Pakistan; we cannot assist Israel without provoking an anti-American outburst from the Arab world; fighting the war in Vietnam enraged the "doves" in many parts of the world. It is clear, then, that a certain price must be paid in at least psychological terms—and it should be incurred knowingly, after both a careful assessment of the benefits and risks and a calculated attempt to devise ways to minimize unfavorable impacts and maximize favorable ones.

After the Tet offensive, Gen William C. Westmoreland said that the results were psychological, not real. Well, he and his colleagues had better learn that when it comes to accomplishing US objectives, psychological factors can be just as "real" as guns, ships, planes, or nuclear weapons. This is because human beings are always the ultimate movers and shakers, and humans are psychological animals.

In short, psychological data needs to be systematically collected and cranked into the intelligence appraisals of given situations. And these data, both domestic and foreign, should be taken into account in framing foreign policies and enunciating international positions. The psychological requirements for achieving US objectives need to be carefully calculated from the start, and every effort must be made, through leadership, persuasion, and public diplomacy, to secure their fulfillment.

Simply put, psychological data is essential if the United States is to be an effective leader in today's world. And its employment is imperative if the president of the United States is going to lead a unified American people into the future.

Notes

1. Lloyd A. Free, "The Role of Public Opinion," *The Forum*, Spring 1971, 55; lecture, National War College, Fort Lesley McNair, Washington, D.C. Reprinted with the concurrence of the National War College and the courtesy of the author.

A Critical Analysis of US PSYOP

Lt Col Philip P. Katz, USA, Retired
Ronald D. McLaurin
Preston S. Abbott

In an era during which it has become very fashionable to discuss the psychological aspects of international relations and during which the study of that field has been greatly influenced by psychological theories and research, particularly in America, little attention has been focused on the communications aspect of international relations; that is, psychological operations. In this essay, we identify and assess the principles of—and some developments in—this field. We also advance several modest proposals for further progress in analysis.

Introduction

Psychological operations is that specialized field of communications that deals with formulating, conceptualizing, and programming goals, and with evaluating government-to-government and government-to-people persuasion techniques. Properly defined, PSYOP is the planned or programmed use of human actions to influence the attitudes and actions of friendly, neutral, and enemy populations that are important to national objectives.[1] The critical variable is, then, the perceptions of foreign populations.[2] Propaganda is only the most obvious example of a persuasive communication. Because psychological operations are designed to influence actions or attitudes, the parameters of the field can be seen in terms of Harold D. Lasswell's classic communications model:

> *Who Says What*
> *In Which Channel*
> *To Whom*
> *With What Effect.*[3]

We know that a government spokesman is saying something in some channel to foreign populations with a view to influencing them. The specific operation is then defined on the basis of what, in which channel, and with what effect. Psychological operations, then, is a truly interdisciplinary phenomenon, lying in the interstices of psychology, sociology, communication, and political and military sciences.[4,5] PSYOP is communication that "embraces the study of persuasiveness, on the one hand, and persuadability on the other. It also involves the study of attitudes—how they are formed and how they can be changed."[6] In practice, however, psychological operations requires inputs from linguists, ethnologists, historians, cultural anthropologists, and area specialists, along with others from the humanities, natural sciences, and social sciences.

Historical Review[7]

Leaders have used persuasive appeals as far back as recorded history allows us to search; Americans, authors of the Declaration of Independence and other brilliant psychological operations, were far from unfamiliar with this tradition.[8] The process of psychological operations remains essentially what it was when Gideon defeated the Midianites but, like many other human enterprises, it has become infinitely more complex. The development of mass communication—broadcasting, worldwide wire news services, mobile printing presses, motion pictures— provides instruments of psychological operations previously undreamed of, as may be seen from the scale on which they were used in World War II as compared with previous wars.

No one who lived through or read about the collapse of France in 1940 will ever forget Goebbels's development and use of psychological operations as major political and military weapons of attack. The Nazis orchestrated the use of radio, the press, demonstrations, group meetings abroad, agents, displays, fifth-column terrorism, and, once the attack started, screaming dive bombers (the memory of which is terrifying to many). The Nazis gave the first full-dress demonstration of what psychological operations can accomplish with the new tools of mass communication and the new weapons of military warfare. And once the meaning of what they had done came to be understood, all the major combatants who had not already done so were compelled to institutionalize psychological operations in their own plans and concepts of modern war. Nevertheless, while there were many programs, no systematic, centralized planning was undertaken in the United States until World War II[9,10]—and the organization and resources developed during the war were quickly reduced following its conclusion. There was some effort to study the lessons learned from PSYOP programs, but the analyses were disseminated unevenly in the government and military services. The civilian Office of War Information (OWI) was disestablished and the propaganda function was passed to the Department of State.

During the 1950s, however, American psychological operations received renewed attention. The highly ideological character of the cold war rivalry and the political nature of the Korean War led to an emphasis on the psychological aspects of conflict. Courses of instruction were developed and "PSYOP" was rationalized within the services. In the civilian government structure, PSYOP was rationalized within the United States Information Agency (USIA). During this period, which was characterized by a political more than a military conflict, channels such as "Radio Free Europe," "Radio Liberation" (later "Radio Liberty"), "Radio in the American Sector" (RIAS), and the "Voice of America" (VOA) were primary PSYOP agents. They communicated news, ideas, and opinions across political boundaries—not infrequently with significant effects.

Unquestionably, the largest American effort in psychological operations since the second World War occurred during the Vietnam War. From the 1950s until the early 1970s, the US government placed substantial emphasis on psychological aspects of its world role, first in its rivalry with the Soviet Union, second in its assumption of global leadership, and finally in Vietnam and other local conflicts. However, the growth of détente and the apparent failure of PSYOP in Vietnam contributed to a continuation of the pattern whereby persuasive communications resources are drastically reduced when there is no longer a perceived immediate need for them.

General Principles

Effective psychological operations requires several basic ingredients: adequate intelligence, coherent organization, sound planning, and a systematic evaluation of feedback. Intelligence is the most commonly overlooked prerequisite to an effective PSYOP program. From a PSYOP perspective, intelligence is the basis for understanding the communications, emotions, attitudes, and behavior of individuals and groups.

Basic to the understanding of persuasive communications is the fact that audience factors are the principal constraint on communication effects. Although studies have been conducted on source attractiveness, trustworthiness, and background, it is the audience's perceptions of credibility that count.[11] Explorations of the phrasing, ordering, and other presentation factors continue to show that the critical variables in determining communication effects are audience factors—education, exposure to opposing views (previous or subsequent to the message), extant opinions, and culture.[12] In addition, such wholly audience considerations as target composition and persuadability are clearly beyond the influence of any communicator.[13]

Since the target population in effect sets the parameters of persuasion, intelligence must provide sufficient accurate data to support the development of programs and messages to optimally affect the audience in all its diversity.[14] Information must be timely, thorough, and systematic.

Organization is important because it must unify the diverse parts of the complex process of intercultural communications. Without sound organization, no single aspect of the process can contribute efficiently to the system. Breakdowns will occur within as well as between program responsibilities. Examples are given below.

Planning is essential at all levels, from tactical operations to long-range, strategic activities. The psychological plan, the basic instrument for integrated and concerted operations, must be designed to exploit all of the source advantages possible. The plan should use intelligence to determine the media, messages, and compositions most likely to support the attainment of national objectives.

To implement PSYOP plans, the program must extend beyond words to actions—and actions must be an extension of words. An effective program of persuasive communications draws on real action attributes and elaborates a realistic and concrete future course. Political and military actions are critical elements in psychological operations.

In order to continue to improve the effectiveness of the program, a system to collect, analyze, and exploit feedback on past operations must be established. Because of the weakness of traditional measures of PSYOP effects, this requirement has too often been neglected.

Organization

It is perhaps a truism to say that an organization should be so constituted that it is capable of effectively carrying out the functions assigned to it. In this section, we emphasize the essential functions of psychological operations and explain how these functions are currently managed in the United States.

An institutionalized structure is essential for the effective implementation of psychological programs. There must be a defined organizational structure with clear lines of authority and responsibility for the development and implementation of policy, plans, and operations. Whether strategic or tactical in scope, the psychological operations mission requires the staff and personnel necessary to develop and implement policy, develop and execute plans, collect intelligence, evaluate programs, and conduct operations.

Policy

A first requirement for conducting meaningful psychological operations is the national-level development and implementation of clearly articulated PSYOP policy in support of political, social, economic, or military goals. The government must establish realistic policy goals to ensure consistent and credible operations. Credibility of policy objectives is the core element in effective psychological programs. False or inconsistent national policy can destroy the best plans and programs.

One example of a poor policy and a lack of understanding was Hitler's attitude toward the Vlasov movement in which he failed to exploit the psychological aspects of Russian Nationalism in 1943.[15,16] Based on policy, appropriate psychological objectives to support operations can be articulated. Programs designed to implement objectives should not be academic—they must be credible and an integral part of the operational environment. For example, if the psychological objective is to induce enemy defection and/or surrender, appropriate actions must be taken by military commanders to ensure that friendly forces clearly understand the surrender policy, will honor safe-conduct leaflets, and will establish appropriate policy concerning the good treatment of enemy prisoners in POW camps. In an insurgency or political environment, the government must develop and implement a credible amnesty policy regarding the insurgents and their supporters.

In the United States, the development and implementation of psychological policy is the responsibility of the executive branch of government. The National Security Council, Department of State (DOS), USIA, and other appropriate departments and agencies form a joint working group or task force to develop psychological policy for an international crisis, event, or a US overseas program. This is an ad hoc group, and policy is determined by consensus.

In peacetime, the US ambassador is responsible for implementing PSYOP policy overseas. In the case of Vietnam, policy was directed from Washington to the US Embassy in Saigon. It was implemented by the Joint United States Public Affairs Office (JUSPAO). More than 60 policy guidance directives were issued by JUSPAO on such diverse topics as "The Use of Prisoners of War in PSYOP Output," "PSYOP Support of Pacification," and "PSYOP Aspects of the Refugee Program." US officials made every effort to develop and promulgate policy directives in conjunction with the Vietnamese government, but such coordination was the exception rather than the rule.

Planning

By definition, a psychological operation is the planned and programmed use of communication media and/or other actions to influence emotions, attitudes, and behaviors of selected target audiences. Planning is the key to psychological operations (random and isolated actions cannot produce consistent results). Planning is essential for psychological operations at all levels—from the strategic and long-range national plan to the PSYOP annex for tactical operations at the battalion command level.

When implemented, the plan becomes the basic instrument and authority for the conduct of psychological operations. One key element in conducting a psychological offensive is the integration of all available assets and their concentration on significant objectives and target groups based on the reality of military, political, or economic operations. The PSYOP plan is the crucial ingredient for integrated and concerted operations. Generally, psychological plans should consist of (1) concept of operations, (2) definition of target groups, (3) clear definition of objectives, (4) general thematic guidance for each objective, (5) injunctions or prohibitions in respect to themes, (6) time-table or schedule to ensure staged and fully orchestrated multimedia operations, and (7) definitive instructions for PSYOP units and assets.

Currently within the US government, there is no central mechanism for the planning of psychological operations. The USIA issues program guidances to its overseas services and divisions. In turn, the United States Information Service (USIS) (overseas) develops country plans which are annual information/cultural programs tailored for a particular country. The Department of Defense (DOD) is responsible for developing psychological plans in support of military contingency operations. Military plans are informally coordinated with the DOS, USIA, and other appropriate agencies.

United States PSYOP planning during the Vietnam period was the primary responsibility of the US Embassy, Saigon. In JUSPAO, the Office of Plans, Policy, and Research was responsible for those plans relating to political, economic, and social programs as well as Republic of Vietnam psychological operations in support of Chieu Hoi (open arms), pacification, and other programs.[17] On the other hand, the US Military Assistance Command, Vietnam (MACV), developed and implemented plans in support of military operations and selected programs directed to target groups in North Vietnam. Generally, PSYOP plans were coordinated with the Republic of Vietnam through its Ministry of Information and/or the Vietnamese Joint Military Command.

Intelligence

Intelligence is the basis for understanding communications, emotions, attitudes, and behaviors of individuals and groups. Y. Tanaka has proposed a three-step strategy of cross-cultural communication that involves quantitative and qualitative analyses of attitude components within a target population, systematic use of a multichannel and multistep flow of communication, and a strategy in which communication is by deed and words.[18] It is significant that the first step in his strategy is the collection of PSYOP intelligence—the one critical step that is often neglected by the political or military planner.

There are three basic requirements for collecting, collating, and analyzing PSYOP data. First, it is essential that managers and media personnel have a thorough understanding of the psychological aspects of the target audience. Anyone engaged in communication and propaganda programs must have certain information about the audience—background, literacy, preferred languages and dialects, art forms and symbols, and key emotional symbols. In addition, PSYOP personnel should have a thorough understanding of the social, political, religious, economic, and military attitudes that are prevalent in significant population groups. With such knowledge, communication programs can be developed to restructure hostile or negative attitudes, reinforce friendly or favorable attitudes, or reinforce/change neutral attitudes, when appropriate.

The second requirement is for timely and systematic operational feedback. Specifically, information is needed regarding the effort being made to support PSYOP objectives, tasks, and themes aimed at definitive target groups. These data are essential for analyzing the success or failure of PSYOP programs and strategy.

The third requirement is for timely and systematic feedback from the target audience when a PSYOP campaign is in progress—a substantial task when the full range of communication media is used. A careful and candid analysis is necessary in order to determine which messages and channels of communication worked well and why, what the mistakes were, and what can be done to avoid future errors and failures.

In 1966, a senior foreign service officer of the USIA listed a number of factors that were inhibiting the effectiveness and measurement of PSYOP programs in Vietnam.

1. We "fly by the seat of our pants" both in setting goals and measuring our success.

2. Our assessment is superficial because we are not experts and because reporting requirements tend to emphasize deeds and statistics and not attitude, opinion, and behavior change.

3. USIA is oriented more to media than to audience.

4. We deal too frequently with accidents rather than essences, which leads to confusion of ends and means.

5. There is a basic lack of research data.

6. We attempt to change rationally attitudes that are basically emotional.

7. We are often confused by changing goals in the field versus our long-range basic mandate.

8. We lack orientation toward communications as a whole and communications research in particular.

9. There is a communications gap between Washington and the field, between media product and field need.[19]

Weaknesses in measuring the effect of USIA programs were stated by another senior officer.

1. The system is too simple in theory and too unclear in practice.

2. Our objectives are very broad, our specific knowledge very narrow. While we can never have all the knowledge we need, we can narrow the gap in some respects.

3. Much basic data on which to base evaluations is missing; e.g., the extent to which messages are directed to

target audiences, whether the messages are received, and whether they are understood.

4. There is at present too little operational research directed at the question of results.

5. Too much evaluation is being done by operators themselves. The evaluation is therefore subjective.

6. In the rush of day-to-day business, the hard questions are often postponed indefinitely.[20]

No single government agency is responsible for the systematic collection of PSYOP intelligence. The Office of Research in the USIA sponsors ad hoc research and conducts some overseas field surveys in order to evaluate selected media output. Intelligence specialists in the Bureau of Intelligence and Research (Department of State) are primarily interested in political information. In a similar vein, the Defense Intelligence Agency is principally interested in the collection and analysis of "hard" military intelligence such as order of battle and weapon systems data. US Army PSYOP intelligence teams generally do not possess the resources or the scope of knowledge required to evaluate the psychological environment of potential target groups around the world.

In Vietnam, as during the Korean War, an attempt was made to overcome PSYOP deficiencies in intelligence by sponsoring ad hoc studies and surveys. However, PSYOP intelligence requirements do not lend themselves to ad hoc or temporary measures: The problem is operational, it is dynamic, and it requires constant attention by professionals who are a permanent part of the team.

A computer-oriented system is being developed by the US military to correct some of the inadequacies in PSYOP intelligence, especially in appraising the success or failure of propaganda programs. This system, known as the PSYOP Automated Management Information System (PAMIS), will be discussed in subsequent sections of this essay.

Operations

Within the US government, the USIA is responsible for educational, cultural, and informational programs and operations overseas. Its operations in Washington (for overseas audiences) include a broadcast service (primarily the "Voice of America"); information center services that support overseas exhibits and information centers; a motion picture and television service that prepares films and audio tapes for release overseas; and a press and publication service that prepares and distributes magazines and other printed matter to overseas posts. Most of USIA's day-to-day operations are the responsibility of the USIS,[21] which is a part of the US overseas mission. The organization and activities of the USIS are dependent on various protocols and agreements between Washington and the host nation.

The US military services have their own PSYOP units and, when authorized by the president, can support USIA operations overseas. Military PSYOP units are composed of various teams that can be put together to accomplish specific missions and functions. US Army PSYOP responsibilities include propaganda development, audio visual production, research and analysis, graphics production, printing, radio operations from mobile studios and transmitters, and loudspeaker operations.

In Vietnam, US military PSYOP favored an organizational arrangement that heavily involved the USIA. JUSPAO was established in Saigon in May 1965, a result of the National Security Council directive. JUSPAO was a PSYOP command supported by USIA, DOD, and USAID (US Agency for International Development). The director of JUSPAO (minister/councillor to the ambassador) issued policy directives (to both military and civilian agencies) and conducted the full range of strategic and tactical operations. At its peak JUSPAO was composed of about 250 US officers, more than half from USIA, and about 600 Vietnamese employees. Within the staff of MACV was the Psychological Operations Directorate, a separate staff section that was later put under the operations chief, the J-3. The military also had operational control of a large PSYOP field unit—the Fourth PSYOP Group. Additional support was available from the Seventh PSYOP Group, which was located on the island of Okinawa.

Programs and Operations

While it is true that Vietnam represented the largest post-World War II American venture into psychological operations, the magnitude and breadth of the US effort there makes Southeast Asia an atypical example of American communications programs—even if a rich source of perspectives.

The major continuing US effort in this field is that of USIA, including the VOA.[22] Through periods of attention to and neglect of military PSYOP, USIA has been the principal[23] channel of American communications to foreign audiences.[24] The USIA was significantly affected by the détente in US-Soviet relations; for example, its once-anticommunist rhetoric has become anachronistic. Its other functions in information dissemination and cultural exchange have withstood the passage of time far better. This is hardly surprising, since persuasive communications have marginal value as opinion changers but are much more effective in attitude reinforcement.[25]

Pre-Vietnam

Specific efforts mounted by US psychological operations before Vietnam include, most prominently, campaigns in Lebanon (1958,[26] 1962[27]) and the Dominican Republic (1965[28]), and American efforts concerning Africa also grew after about 1960. Throughout the entire period, "Radio Free Europe,"[29] "Radio Liberty,"[30] "Radio in the American Sector" (Berlin),[31] and other propaganda outlets were directed toward the USSR and Soviet-dominated areas.

The evolution of the postwar world left the United States, notwithstanding its early history and national values, in the position of the world's most important status quo power.[32] American international communications were therefore generally antirevolutionary, frequently supporting unpopular regimes in the name of stability.[33] In some cases (e.g., Lebanon) the programs were reasonably effective, but psychological effort success is difficult to separate from the nature and success of the overall political or military activity.

Vietnam

The importance and priority that the North Vietnamese and Vietcong (VC) put on psychological operations are well known, as in the slogans "Political activities are more important than military activities," and "Fighting is less important than propaganda." Vo Nguyen Giap in his *People's War, People's Army* quotes as one of Ho Chi Minh's cardinal principles of political warfare, "Do not attempt to overthrow the enemy but try to win over and make use of him."

As noted above, in 1965 JUSPAO was established to coordinate the US policies and personnel involved in psychological operations. A major portion of the new US PSYOP effort was to be devoted to the Chieu Hoi returnee program (an attempt to win over and make use of the VC). There was to be optimum coordination and integration of both US and Vietnamese operations at all levels, with overall supervision from JUSPAO and the Republic of Vietnam Ministry of Information (MOI) through the Vietnam Information Services vested in a combined US-Vietnamese coordinating committee at the national level with representation from MACV, JUSPAO, General Political Warfare Directorate (GPWD), and MOI.

"Guidelines to Chieu Hoi Psychological Operations: The Chieu Hoi Inducement Program" was prepared in April 1966 by JUSPAO. It centralized policy planning and decentralized operational planning. Execution was assigned to the local level. Development and mass production of PSYOP materials were done by JUSPAO in accordance with tactical needs determined by the field. In addition to JUSPAO, MACV (the 4th US Army PSYOP Group) operated an extensive program, much of which was devoted to the Chieu Hoi inducement program.[34, 35]

Content of PSYOP material, targeted on the potential rallier, was focused on his grievances, emotions, and aspirations—not on ideological commitment (except in the case of hard-core VC). The insurgent was encouraged to return to his home by the creation of trust in the government as just and generous. Former insurgents were used in preparation of the material to the maximum extent possible—they, rather than the Americans, knew the modus operandi of the enemy and were a part of the indigenous culture.

The leaflet—distributed from aircraft and by hand—proved to be the most practical means of disseminating the Chieu Hoi message. The ubiquitous "safe conduct pass," which literally blanketed South Vietnam, was probably the most effective message. Though there were thousands of other leaflets stressing many other themes,[36] the safe conduct pass was most often described by ralliers during interrogation as the one most seen and the one most conducive to rallying. After one battle, 90 percent of those VC who could be searched—the dead, wounded, and captured—had the safe conduct leaflet.[37] By the spring of 1971, JUSPAO had distributed nearly four billion leaflets in the campaign to persuade "men to rally to the GVN [Government of the Republic of Vietnam] under its amnesty program."[38]

No discussion of the factors motivating defection is complete without mention of the carefully structured program of rewards given to those who rallied and indeed even to those who influenced a VC to rally. The Chieu Hoi weapons reward program paid a returnee for weapons he turned in or for weapons that he was able to recover after he rallied. This program was successful in locating large weapons caches. The weapons reward system was established by official decree of the Chieu Hoi Ministry of the GVN.

The Vietcong military forces neutralized through the Chieu Hoi program were equal to about one-fifth of the total of all their forces killed or captured by military action. More than 4,000 men from the government of the Republic of Vietnam or its allies would have been required to effect the same result through military action.[39]

Besides the safe conduct leaflet and rewards program, testimonials from Hoi Chanh (former VC) proved to be effective in the total psychological process of attitude change. It was determined that the Hoi Chanh testimonial should contain four essential elements: a photograph and complete individual description of the Hoi Chanh; an indication of why he rallied; discussion of the good treatment he received; and an appeal to his former comrades to rally. Experience in the field showed that when a PSYOP person wrote a testimonial message for a Hoi Chanh, it was usually recognized as propaganda. The best approach was for the Hoi Chanh to address his message specifically to his former unit and address some of his former comrades by name. He should tell enough about himself to convince the recipients of the message that he is in fact alive and well. The operative word in all Hoi Chanh testimonials was credibility. Themes were suggested to the returnee, but the language was his own. In many instances, audio tapes were prepared of Hoi Chanh testimonials for broadcast by air-to-ground loudspeaker systems and radio.

Poetry, music, and art were effective PSYOP media directed to hostile as well as friendly target groups. The messages were emotional in tone and concerned such time-tested themes as love, family, religion, home, holidays, and national heroes.[40]

It became apparent as a result of operational experience that direct psychological operations, aimed at modifying attitudes and behavior, must be preceded by propaganda that is indirect in character, slow and general in nature. It should seek to create an atmosphere of favorable preliminary attitudes. No direct psychological operations can be effective without a proper foundation. Safe conduct passes and testimonials are one type of foundation; that is, they inform the enemy soldier that if and when he decides to surrender, certain protocol will be observed. The best psychological operations involve face-to-face persuasion. The following example of PSYOP messages being reinforced by face-to-face persuasion was provided by a former deputy leader of a VC surgical team who was also a Communist party member.

> The Chieu Hoi leaflets were dropped by airplanes very often. I read some of them . . . At that time I didn't really believe in the leaflets because I was told by the cadres that the government wanted to cheat Front people by its Chieu Hoi program. Although I didn't believe what the leaflets said, I had something in my mind about leaving the Front. When my friend, Mr. Thiet, came to tell me about the government's treatment toward the ralliers, I believed him. I believed that I wouldn't be beaten up by the GVN as the cadres had said, because my friend was one of the ralliers, and he wasn't beaten up or punched down into the sea. [This refers to VC claims that GVN troops took the Hoi Chanh up in a helicopter and then threw them into the ocean.] On the contrary, he was well-treated by the government. He was given a job, and he was free to visit his family when he wanted.[41]

Another type of foundation used in an insurgency environment was a magazine disseminated in Vietnam called *Long Me.*

135

The primary target audience for the magazine was the families, friends, and sympathizers of the VC. The 62-page, full-color magazine was distributed bimonthly to villages and districts known to be sympathetic to the VC. It was distributed by priests, merchants, school teachers, bus drivers, fortune tellers, armed propaganda teams, and others. *Long Me* was also distributed through direct mail. It contained feature stories, poems, short novels, cartoons, and art. The feature stories were factual, and they emphasized the positive features of the new life offered to former VC. *Long Me's* indirect approach worked well. The staff of *Long Me*, with the exception of one American (Philip Katz), was composed of former communist propagandists.

Not all propagandists in Vietnam believed in the slow and indirect approach. Many messages were harsh, explicit, and direct; for example, this leaflet was prepared and disseminated by an American military unit:

> The sky soldiers of the US 173d Airborne Brigade are here to destroy you. We have killed hundreds of your comrades with our powerful bombs and artillery and invincible ground troops. You will die a senseless death and be buried in an unmarked grave in an unknown jungle. Rally now before it is too late.

Propaganda must be continuous and lasting—continuous in that it must not leave any gaps in logic and continuity of themes, lasting in that it must function over a period of time. The principle of repetition plays an important role in persuasive communication.

PSYOP is not magic; effective communication is based on slow and constant impregnation. PSYOP is not a stimulus that disappears quickly; it consists of successive messages and actions aimed at various emotions or thoughts through many instruments of communication. Effective psychological actions should be continuous; as soon as the effect of one message is weakened, it should be reinforced by another.

Post-Vietnam

After the conflict in Southeast Asia, the United States once again moved away from systematic, coordinated psychological operations on a government-wide scale. As in the past, military human resources for PSYOP were reassigned or neglected. The USIA had no intensive focus, and USIA efforts were only marginally coordinated with other US persuasion programs. The lessons of the past concerning the potential value of PSYOP, the importance of coordinated, systematic planning, and the value of continuous communications as facilitator (even during periods of strategic security) were once again ignored. [42]

Effectiveness

Probably the most elusive single area in the analysis of psychological operations is the evaluation of PSYOP effectiveness. How can effectiveness be measured? Polling dominates much of our political life, and market research is considered fundamental to sound business practices; yet PSYOP operates in an anachronistic world where direct audience analysis is often impossible.

Effectiveness of persuasive communications is generally viewed in terms of the degree to which attitudes are changed or reinforced. However, studies have conclusively demonstrated that PSYOP directed at denied or hostile areas should aspire to modify peripheral attitudes only, not fundamental ones.[43] Our government's persuasive communications to foreign audiences enjoy very limited parameters of only potential effect.[44]

The relatively small range of impact and the difficulty of measurement are primary constraints in the evaluation of effectiveness. The fundamental question is, how successful are the communications in terms of the limited objectives? Sometimes, however, practitioners have been forced into a secondary question: Assuming the general principles we have identified to be optimum, to what degree are the communications congruent with them?

Another major impediment to effectiveness assessment is the nonpersuasive nature of psychological operations. One is inclined to think in terms of the critical message directed at the critical audience at the critical moment. In fact, the vast majority of propaganda and other instruments of PSYOP consist of what L. John Martin has called facilitative communication (*not* persuasive communication).[45] That is, PSYOP is designed to keep the lines open—maintain the channel and its credibility—in *preparation* for the critical movement, message, and audience. Since the purpose of the messages sent in facilitative communication is *not* persuasive, how do we measure their effectiveness? To what extent should the two types of communication, which are purposively related but conceptually discrete, be evaluated by a single measure?

> If we could select our audience on the basis of certain idiocratic factors—objective physical and personal characteristics peculiar to an individual, such as age, sex, race, education—we might increase by a statistically significant fraction the proportion of those influenced by a message. But we would have no control over such factors as personality and susceptibility to persuasion, existing values, beliefs and opinions or attitudes toward the objects, subjects and situations involved in the persuasive message. We can choose our communicator but not determine his image. We can select the vehicle of transmission but not the channel of reception of the target of our communication. We could maximize the effect of all these factors for a single individual, especially if we were able to subject him to intensive precommunication analysis. But there is no way that this can be done for the diverse assortment of individuals who normally make up the audience of the mass media, the vehicles most commonly used in international propaganda.[46]

For lack of a criterion,[47] as well as for the inability to secure direct access to the audience and the recognition of the inherent limits of persuasive communications, the most common evaluations of effectiveness have been output measures, despite their notorious limitations.[48] A second approach has been to focus on reception; in general, this area has not attracted much attention. The final type of analysis—effects measures—may be both direct and indirect.

Direct measures of effects can involve the use of polling or survey data. In peacetime, such information can be gathered in interviews with traveling foreign nationals,[49] direct mail questionnaires, and polling in the audience's country. In war, or when relations between the sender (source) and receiver (audience) countries are strained, only some—or none—of these may be practicable. In these cases, refugees[50] and prisoners of war[51] have been used to gather effects data. However, such groups often have inherent disadvantages as a population—they may not be a representative sample, and identification of the role of any individual message in their decision to defect is usually elusive.

Indirect measures include the content analysis of intercepted mail, mail drops,[52] captured documents, broadcasts and telecasts, and target area public media.[53]

Trends

In an effort to institutionalize PSYOP intelligence, the US military designed and developed a computer-oriented system called PAMIS. This information system was designed to fulfill several objectives: (1) to provide the spectrum of information needed to support PSYOP organizational elements for the planning, implementation, and evaluation of PSYOP programs; (2) to encourage (indeed, to enforce) methodologies to support information-gathering programs; and (3) to provide an appropriate automatic data-processing system for the storage, analysis, and utilization of the gathered data.

PAMIS encompasses these three systems:

> Foreign Media Analysis (FMA) System,[54]
> PSYOP Foreign Area Data System (PFAD),[55] and
> PSYOP Effects Analysis System (PEAS).[56]

The primary objective of FMA is to provide statistical data obtained from publications and radio broadcasts of selected foreign countries. The data collected includes information about their propaganda objectives and strategy; propaganda themes used to support these objectives; propaganda trends; the tone and intensity of media coverage related to domestic themes, foreign governments, international organizations or political movements; and propaganda techniques and tactics. Another FMA objective is to provide clear-text abstracts on important events, subjects, and persons as they are treated in the public media of the selected countries.

The objective of PFAD is to provide the US military PSYOP community with uniform methods and procedures for the collection, storage, and rapid retrieval of data needed for planning and operations. The PFAD was designed to accomplish the following tasks:

- define parameters for the collection of PSYOP-relevant data;
- provide uniform procedures and methods for collecting and reporting such data;
- provide a retrieval system that has a centralized data bank, the core of which is a set of over 1,600 foreign area data descriptors, and that permits retrieval of two types of reports: annotated bibliographic material and narrative summaries of subject data; and
- provide for continuous and rapid updating of PSYOP foreign area information.

The computer system provides the PSYOP community with social, cultural, economic, political, communication, and other information that is needed to make decisions on aspects of PSYOP policy, planning, allocation, and deployment of resources and operations.

PEAS was designed to collect, store, and process data from a sample of a target audience in order to measure the success or failure of PSYOP messages and programs. The system also stores and processes data about friendly and hostile PSYOP activities and actions. Timely, accurate reporting on the purpose (objectives and tasks) of friendly PSYOP programs is essential, as are facts about the distribution and intensity of efforts expended on specific target audiences. Although hostile PSYOP activities and actions usually do not directly affect friendly messages, systematic monitoring of hostile propaganda and other actions provides analysts with many clues about how enemy elites react to our PSYOP programs. In fact, when respondents from the target population (such as prisoners, defectors, and refugees) are not available, the content analysis of hostile media can be the most valuable source of feedback concerning the impact of friendly programs.

A significant objective of PEAS is to institutionalize methods and procedures for analyzing PSYOP programs and strategies so that the successes and failures of strategic and tactical programs can be measured. PEAS was therefore designed to provide the following:

- uniform (institutionalized) procedures and methods for reporting and describing all PSYOP activities in a theater of operations;
- rapid retrieval of command and control data to determine how effort and cost relate to objectives and tasks;
- the data necessary for analysts to correlate operational data with specific military target groups and segments of the civilian population;
- data for a continuous evaluation of the initial impact of PSYOP messages (audio and print) directed to general or particular target groups;
- uniform procedures and methods for reporting and analyzing hostile propaganda and action directed to friendly military and civilian populations; and
- data needed for an independent audit and assessment of military PSYOP by the joint or other appropriate command.

To accomplish the above, PEAS was designed to accept, store, and process data inputs into three computer files.

The ACTIV file of PEAS was designed to institutionalize the reporting of military psychological activities at each command level. This reporting system provides commanders with systematically collected data for command and control of the full range of PSYOP activities. Specifically, the ACTIV file provides information about the quantity of effort being directed in support of specific programs or campaigns, in support of PSYOP objectives, tasks, and themes, and toward definitive target groups. In addition, this file provides information about the amount of effort allocated to each channel of communication (such as print, radio, and loudspeaker). Also available is information about psychological actions such as show of force, exploitation of new weapons, and supply drops.

Much basic data is needed for making evaluations: To what extent are our messages directed to specific target audiences? What are our objectives? Are our messages received and understood? PEAS will provide many answers to these questions.

The IMPACT file of PEAS was designed to institutionalize the collection of data about the effect of specified propaganda messages, activities, and programs. The data collection format was designed to be manageable for prisoner of war or other interrogations at any level of command. Each IMPACT computer record pertains to a respondent's (such as POW) reaction to an electronic or printed message or action. In the first step (of the interrogation process), the respondent is required to select from a catalogue of PSYOP messages that he has seen or heard. The IMPACT data record is then completed for those messages identified by the respondent.

It is beyond the scope of this essay to present the details for the multitude of inputs and outputs of the ACTIV file of PEAS, but both quantitative (statistical) and qualitative (textual) reports are available to analysts and other users of the system. These reports provide gross information related to four indicators of PSYOP impact.

- *Understanding:* Did the audience understand the language and symbolism used in the message?
- *Credibility:* Was the content credible, or did the audience view the message as obvious propaganda?
- *Influence:* Did the message influence the respondent to act in a way favorable to the sponsor of the message?
- *Irritation:* Did the language, symbolism, or other aspect of the message irritate the audience?

Statistical analysis of such data should reveal the specific themes and messages that were successful; it should also reveal those messages that failed. In addition, PEAS should provide analysts with a generous number of textual responses, in the respondent's own words, that explain the reasons for both success and failure.

PEAS will provide military commanders with the range of information needed to evaluate the impact of psychological operations. No longer will number of leaflets produced or hours of broadcast time be the only—or even the basic—criteria for evaluating PSYOP programs.

The POLWAR file of PEAS provides uniform procedures and methods for reporting and analyzing hostile propaganda and action directed to US military—and other friendly—populations. Data from this file allow for the continuous monitoring of hostile POLWAR activities at the tactical command level. It is this file of the PEAS that interfaces with the FMA system. The

POLWAR file is concerned with those communications, propaganda, and actions that take place in a combat environment and are directed at friendly military and civilian audiences. This contrasts with the FMA system, which is concerned primarily with strategic objectives and which deals with data obtained from newspapers, magazines, and radio broadcasts directed to their own domestic audiences.

Conclusions

Tanaka's concept of cross-cultural communication, noted above, emphasized the systematic use of multichannel, multistep flow of communication, and a strategy in which communication is by both deed and word.

Propaganda must be total. A modern psychological operation must utilize all of the technical means at its disposal—the press, radio, TV, movies, posters, meetings, and face-to-face persuasion. There is no psychological operation when meetings and lectures are sporadic, a few slogans are splashed on walls, radio and television presentations are uncoordinated, and news articles are random. Each communication medium has its own particular mode of influence—alone, it cannot attack individuals, break down their resistance, or make their decisions for them. A film does not play on the same motives, does not produce the same feelings, does not provoke the same reactions as a leaflet. The very fact that the effectiveness of each medium is limited to one particular area clearly shows the necessity of complementing it with other media. A word spoken on the radio is not the same as the identical word spoken in private conversation or in a public speech. Nor does a word appearing in print produce the same effect as the same word when it is spoken. To draw the individual into the net of persuasion, each technique must be utilized in its own specific way, directed toward producing the optimum effect, and orchestrated with all the other media. Each medium reaches the individual in a specific fashion and makes him react anew to the same theme—in the same direction, but differently. Thus, no part of the intellectual or emotional personality is left alone.

Effective totalitarian regimes have a clear understanding of the psychological aspects of military actions. These regimes employ propaganda for psychological impact every day, and propaganda is fully institutionalized within the political and military structure of these states. Furthermore, these regimes are sensitive to the psychological environment of their target audiences and are able to orchestrate the various instruments of communication to promote unity in their own ranks and unity with their own people while promoting divisiveness in enemy forces.

As a democracy, the United States has tried to renounce domestic propaganda; as a nation in a highly competitive political world, America must use psychological operations. As yet, however, the formula that would support both these aims without their working at cross-purposes has not been found—a fact made quite evident by the awkwardness that characterizes US PSYOP efforts at home and abroad.

Psychological operations can be among the most suitable and effective weapons in pursuit of US national objectives in this unstable and sometimes threatening world. PSYOP exploits military readiness already required for other purposes without requiring the use of military force. It draws power from the strength of Western democracy and the universals of civilization. It can encourage masses of people to develop untapped natural resources and to oppose totalitarian regimes.

The requirements for effective psychological operations are considerable. Modern persuasive communication is not a haphazard, hit-or-miss activity but a well-coordinated, sustained effort. It demands a broad and thorough knowledge of a wide variety of psychological, sociological, and cultural factors. It also requires planning well in advance of events. It must, at all times, serve under an appropriate form of centralized direction with trained personnel and dedicated resources. Contemporary psychological operations is not an amateur's game.

Notes

1. This definition is adapted from Joint Pub 1-02, *Department of Defense Dictionary of Military and Associated Terms*, 1974. *Psychological operations* is a broader term than *psychological warfare* in that it deals with friendly and neutral, as well as hostile, audiences.

2. See R. J. Barret, "PSYOP: What Is It?" *Military Review* 42 (1972): 57–72. For an excellent study of the dynamics of international perception, see Robert A. Jervis, *Perception and Misperception in International Politics* (Princeton, N.J.: Princeton University Press, 1976).

3. Harold D. Lasswell, "The Structure and Function of Communication in Society," in L. Bryson, ed., *The Communication of Ideas* (New York: Harper and Row, 1948).

4. Indeed, the four principal figures in the establishment of communications research—the fundamental PSYOP root—were two psychologists (Paul Lazarsfeld and Carl Horland), a sociologist (Kurt Lewin), and a political scientist (Harold Lasswell).

5. Wilbur Schramm, "Communication Research in the United States," in idem, ed., *The Science of Human Communication* (New York: Basic Books, 1963), 1–16.

6. Y. Tanaka, "Psychological Factors in International Persuasion," *The Annals of the American Academy of Political and Social Science* 398 (1971): 50–54.

7. Historical literature dealing with US PSYOP is far from rich. It includes the following: W. E. Daugherty and Morris Janowitz, eds. *A Psychological Warfare Casebook* (Baltimore, Md.: Johns Hopkins University Press, 1958); M. Dyer, *The Weapon on the Wall* (Baltimore, Md.: Johns Hopkins University Press, 1959); Harold D. Lasswell, *Propaganda Technique in World War I* (Cambridge, Mass.: MIT Press, 1971); D. Lerner, *Sykewar* (New York: George W. Stewart, 1951); and P. M. A. Linebarger, *Psychological Warfare* (Washington, D.C.: Combat Forces Press, 1954).

8. H. Butterfield, "Psychological Warfare in 1776: The Jefferson-Franklin Plan to Cause Hessian Desertions," *Proceedings of the American Philosophical Society* 94 (1950): 233–41.

9. Relatively little has been written in recent years about the so-called Creel Committee, the Committee on Public Information, created by United States President Woodrow Wilson after the United States entered World War I. Substantial writing on the committee appeared in the interwar period. Even about the Second World War it could be said that "U.S. propaganda agencies created during [the war] were more the result of trial and error planning and ad hoc improvisations than of careful blueprinting."

10. Daugherty and Janowitz, 126.

11. C. I. Hovland, I. L. Janis, and H. H. Kelley, *Communication and Persuasion* (New Haven, Conn.: Yale University Press, 1953); C. W. Sherif, M. Sherif, and R. E. Nebergall, *Attitude and Attitude Change* (Philadelphia, Pa.: W. B. Saunder, 1965); C. I. Hovland and I. L. Janis, *Personality and Persuasibility* (New Haven, Conn.: Yale University Press, 1959); R. L. Rosnow and E. J. Robinson, eds., *Experiments in Persuasion* (New York: Academic Press, 1967).

12. C. I. Hovland, *The Order of Presentation in Persuasion* (New Haven, Conn.: Yale University Press, 1957).

13. Hovland and Janis.

14. Tanaka, 50–54.

15. In the summer of 1942 Russian Lt Gen Andrei A. Vlasov was captured by the Germans. The Germans soon learned that Vlasov wanted to lead a Russian national army of liberation into the Soviet Union. General Vlasov (a Russian military hero) was willing to collaborate with the Germans for the sake of Russian national independence. Hitler, however, saw Vlasov at most as a tactical propaganda tool to weaken the Soviet forces: "promises might be given to him and deserters who came over to join him (the Vlasov movement), but on no account should these promises be kept."

16. R. Gehlen, *The Service: The Memoirs of General Reinhard Gehlen* (New York: The World Publishing Company, 1972), 84.

17. The Chieu Hoi program was the campaign to encourage "ralliers" (Hoi Chanh) to come over to the Republic of Vietnam side. Only South Vietnamese (VC) were eligible. Substantial publicity was given to the act of "rallying"— to the motivations and aspirations of the Hoi Chanh, the benefits of rallying, and the like.

18. Tanaka.

19. US Information Agency, *Proceedings of the Seminar on Effectiveness*, R-121-66 (Washington, D.C.: Office of Policy and Research, 1966).

20. Ibid.

21. The *agency* responsible for programs and operations is USIA; the overseas *posts* are USIS posts, and they form an integral part of the US country team headed in each country by the chief of mission, usually an ambassador.

22. There have been many proposals over the years to sever VOA from USIA, but for the present the tie remains.

23. Other US institutions principally involved on a continuing basis include the National Security Council, the DOS, the Agency for International Development, and the DOD.

24. A major source on USIA is J. Henderson, *The United States Information Agency* (New York: Praeger, 1969); M. G. Lawson et al., *The United States Information Agency during the Administration of Lyndon B. Johnson, November 1963–January 1969* (Washington, D.C.: Government Printing Office [GPO], 1968) is an administration history of some value in understanding USIA. T. Sorenson, *The Word War* (New York: Harper & Brothers, 1968) gives an excellent overview of the problems and competing principles with which such an agency must deal. A more critical, briefer, but equally stimulating presentation is K. R. Sparks, "Selling Uncle Sam in the Seventies," *The Annals of the American Academy of Political and Social Science* 398 (1971): 113–23.

25. J. T. Klapper, *The Effects of Mass Communication* (Glencoe, Ill.: Free Press, 1960).

26. B. H. Cooper, Jr., "The Undenounced Intervention," in Ralph D. McLaurin et al., ed., *The Art and Science of Psychological Operations* (Washington, D.C.: GPO, 1976), 241–46; Idem, "Accentuating the Positive," in ibid., 280–86.

27. R. Nathan, "Cuba: Strategic Dilemma," in ibid., 259–61; B. H. Cooper, Jr., "Effective Diplomacy: An Exit from Armageddon," in ibid., 386–91.

28. B. H. Cooper, Jr., "Teamwork in Santo Domingo," in ibid., 229–32; Idem, "Divided Counsels," in ibid., 262–66.

29. R. T. Holt, *Radio Free Europe* (Minneapolis: University of Minnesota Press, 1958); A. A. Mitchie, *Voices through the Iron Curtain: The Radio Free Europe Story* (New York: Dodd, Mead, and Company, 1963); and J. R. Price, *Radio Free Europe—A Survey and Analysis* (Washington, D.C.: Congressional Research Services, 1971).

30. J. C. Whelan, *Radio Liberty—A Study of Its Origins, Structure, Policy, Programming, and Effectiveness* (Washington, D.C.: Congressional Research Services, 1972).

31. E. Taylor, "RIAS: The Voice East Germany Believes," *The Reporter*, 10 November 1953, 28–32.

32. R. L. Heilbroner, "Counterrevolutionary America," *Commentary* 43 (1967): 31–38.

33. G. V. Allen, "What the U.S. Information Program Cannot Do," in J. B. Whitton, ed., *Propaganda and the Cold War* (Washington, D.C.: Public Affairs Press, 1963).

34. The Chieu Hoi inducement program was a campaign to reward Hoi Chanh who turned over weapons. (The reward was money.)

35. J. A. Koch, *The Chieu Hoi Program in South Vietnam (1963–71)* (Washington, D.C.: DOD, Advanced Research Projects Agency, R-1172-ARPA, January 1973).

36. See Philip P. Katz, *Communicating with the Vietnamese through Leaflets* (Saigon: Joint US Public Affairs Office, 1968).

37. Ibid., 66.

38. Ibid., 68.

39. Ibid.

40. Philip P. Katz, *A Systematic Approach to PSYOP Information* (Kensington, Md.: American Institutes for Research, 1970).

41. Joint US Public Affairs Office, *PSYOP in Vietnam: Indications of Effectiveness* (Saigon: JUSPAO Planning Office, 1967), 120–21.

42. See L. A. Free, VOA Language Priority Study, Washington, D.C., US Information Agency, 1969.

43. D. Lerner, "Is International Persuasion Sociologically Feasible?" *The Annals of the American Academy of Political and Social Science* 398 (1971): 41–49.

44. D. D. Smith, "Some Effects of Radio Moscow's North American Broadcasts," *Public Opinion Quarterly* 34 (1970–71): 539–51.

45. L. John Martin, "Effectiveness of International Propaganda," *The Annals of the American Academy of Political and Social Science* 398 (1971): 61–70.

46. Ibid.

47. D. D. Robinson, *A Brief Review Study of the Problems of Criteria in Psychological Warfare* (Columbus, Ohio: Battelle Memorial Institute, Remote Area Conflict Information Center, 1967).

48. Daugherty and Janowitz, 681–84.

49. L. B. Szalay, "Audience Analysis," *Congressional Record* 148 (6 March 1972): 53426-28.

50. Whelan.

51. Philip P. Katz, "A Survey of PSYOP Intelligence," in Ronald D. McLaurin et al., ed., *The Art and Science of Psychological Operations*, 478–94.

52. Mail drops for these purposes are addresses purportedly for give-aways or other contacts. All mail arriving at these addresses is then analyzed for background data—location, age, and sex of correspondent.

53. McLaurin, chaps. 8 and 9.

54. Philip P. Katz, PSYOP *Automated Management Information System (PAMIS): Foreign Media Analysis* (Kensington, Md.: American Institutes for Research, 1972).

55. Philip P. Katz, *PSYOP Automated Management Information System (PAMIS): Foreign Area Data Subsystem (PFADS)* (Kensington, Md.: American Institutes for Research, 1973).

56. Philip P. Katz, *PSYOP Automated Management Information System (PAMIS): PSYOP Effects Analysis System* (Kensington, Md.: American Institutes for Research, 1976).

PART III

Strategic, Tactical, and Operational PSYOP

Introduction

This section deals with the objectives and activities of strategic, tactical, operational, and other types of PSYOP. In all forms, PSYOP primarily supports the attainment of national policy and objectives. The key to all US PSYOP is credibility of the message. Evaluation is based on changing perceptions, attitudes, and behaviors.

Certainly, the ideology of a government plays a role in its PSYOP methodology and its international communications. The authors offer articles about Soviet PSYOP campaigns and activities with a credibility/disinformation perspective.

DeWitt S. Copp discusses Soviet active propaganda measures, including forgery, agents of influence, and disinformation. His discussion serves as a historical summary of Soviet campaigns and ideology.

James Melnich emphasizes mirror imaging in Soviet propaganda. His essay is useful in understanding the overall Soviet PSYOP threat and possible related weaknesses.

Lev Yudovich reveals the significant threat and impact of the Soviet military doctrine of cultivating hate in their soldiers. This doctrine was designed to make Soviet soldiers psychologically prepared to act in dangerous situations and to believe in victory.

Dr Joseph S. Gordon focuses on research, target analysis, pretesting and effects analysis, and propaganda analysis. He concludes that intelligence activities are critical to PSYOP.

George V. Allen, former director of USIA, states that we must be realists and understand that a combination of political, economic, psychological, and military efforts are necessary. He believes that propaganda alone in the short run can do little to solve problems. Allen advocates an honest, objective, and truthful information program that faithfully relies on the individual to make right decisions.

Lt Col John Ozaki, USA, addresses the planning and coordination of successful defector programs as another form of combat power. He emphasizes the characteristics and techniques of a sound program to defeat insurgency movements.

Ronald D. McLaurin emphasizes that PSYOP at the tactical level can contribute to the major strategic objectives of a government and a reacquisition of the loyalty and support of the population.

Benjamin F. Findley, Jr., reviews and analyzes selected US and Vietcong PSYOP in the Vietnam War.

Soviet Active Measures

DeWitt S. Copp

Although Vladimir I. Lenin did not coin the term *active measures* (*activnee meropriyatia*), he originated and set in motion all its component parts. Those parts included the International Department (ID) of the Central Committee of the Communist Party of the Soviet Union (CPSU), Section A of the first chief directorate of the KGB, and the propaganda department (PD) of the Central Committee.

The ID is the coordinating center for Soviet active measures. Its lineage can be traced directly to Lenin's formation of the Comintern in 1919. Lenin's purpose for the Comintern, or Third International, was to advance the cause of global revolution by forming communist parties in noncommunist countries and creating conditions for revolt through propaganda and agitation. The approach was adaptable to political realities.

The International Department, formed in 1957, was a sophisticated apparatus at the summit of Soviet power. It carried the view of the leadership, bringing influence to bear on any area—political, military, or economic—that would aid or abet Soviet policy. At the same time, it sought to denigrate and undermine the policies of the noncommunist world, particularly those of the United States. More politically realistic than its Leninist ancestor, the ID's purpose was to manipulate public opinion and gain the acceptance of Soviet ends. The ID's secretary was Anatolii F. Dobrynin, who had served as Soviet ambassador to the United States for 24 years. He was appointed to his post by Soviet General Secretary Mikhail S. Gorbachev in June 1986, several months after the important Twenty-Seventh Congress of the CPSU.

The ID worked closely with the propaganda department, headed by Party Secretary Aleksander N. Yakovlev, who had served for 10 years as Soviet ambassador to Canada.

The propaganda department employed dual courses of action. It orchestrated the line of the party, utilizing Soviet and bloc media, and overtly replayed disinformation generated by the active measures apparatus. Since the PD embodied the former international information department of the Central Committee of the CPSU, it covered both the Soviet home front and the foreign scene.

One of Lenin's first acts on seizing power was to establish the Cheka, or secret police. He knew that without such an instrument of fear, his movement could not expect to survive. That circumstance did not change, but the Cheka became the KGB—and its activities were worldwide!

Section A of the KGB's first chief directorate worked in liaison with the ID in carrying out covert Soviet active measures. These efforts included disinformation operations, the placement of forgeries, and the managing and directing of knowing and unknowing agents of influence—professionals in sensitive positions, both private and public—whose activities were frequently of great value in the Politburo's plan to gain acceptance of its position on major issues.

Together, the ID, the KGB, and the PD were the architects and planners of Soviet active measures. With an estimated annual budget of $4 billion, their combined efforts were a major force in the mechanics of Soviet foreign policy.

Stanislav Levchenko, a former major in the KGB, was acting director of line PR of the Soviet active measures group in Tokyo when he defected to the United States in 1979. In his 13–14 July 1982 testimony at the Hearings before the Permanent Select Committee on Intelligence, House of Representatives, he gave the following appraisal.

The size of the overt and covert active measures is massive . . . I can tell you from an insider's vantage point that the ID and the KGB receive all the resources and personnel needed to carry out this massive effort. There are never any shortages. Of course, this is not a recent development. An examination of the history of the CPSU will demonstrate the importance of such tactics . . . all active measures operations are assessed against a set of standards. Success is a vital ingredient. The growth of the Soviet active measures effort over the last five to ten years is due to progress in the field.

Forgeries

Soviet active measures were carried out within a spectrum of white (overt), gray (semicovert), and black (covert) operations. It was in this last area that the crafting and placement of forgeries was conducted by section A of the first chief directorate of KGB. KGB's management of forgeries was global, and was linked to the use of Soviet assets in whatever locale the particular forgery was to be surfaced. The KGB's purpose in placing forgeries was to undermine US relations with allies and nonaligned governments and to influence public opinion against US policies concerning sensitive issues. An additional purpose was to convince those affected by the impact of the forgery that decisions announced by the United States cannot be believed because underlying them were secret plans that were harmful in their implication and intent.

For example, an obscure French-language weekly newspaper in Madagascar—*Carrefour* (Crossroads)—carried on its front page (summer 1985) a purported letter from a US Army medical doctor at Walter Reed Research Institute written to a supposed superior at the Pentagon. The substance of the forgery indicated that US and South African scientists were secretly working on psychotropic chemical weapons and so-called ethnic weapons that would kill only black-skinned people.

The forgery contained numerous technical errors, and its falsity was quickly exposed by the United States Information Agency (USIA). However, this did not stop Tass (the Soviet official news agency) at year's end from running the forgery as a true account. Tass lifted it from *Krasnaya Zvezda* (Red Star), the official publication of the Soviet Ministry of Defense, under the headline "Sinister Plans."

The purpose of the forgery was to instill and spread the belief that the US government is vicious, racist, and scientifically willing to pursue inhuman ends. That KGB effort to arouse black Africa failed—the forgery was not picked up by other African newspapers and outlets. However, one of the basic consistencies of the active measures apparatus was that of repetition. A lie once told is never discarded, no matter how thoroughly exposed. In June 1987, in a meeting with USIA Director Charles Wick, who was in Moscow for the opening of a USIA cultural exhibit, the director of Novosti repeated the ethnic weapon lie. Wick walked out of the meeting. Novosti, a supposedly independent Soviet news agency, is heavily staffed by the KGB.

The KGB made a practice of attacking US scientific and medical activities on a broad scale, employing not only forgeries but also disinformation by clandestine radio stations. One such KGB effort was mounted through "Radio Halgan," operated by a communist-front group of Somali dissidents in Addis Ababa, Ethiopia. The station reported that the president of Somalia was negotiating with the US commander of Central Command to lease Somali territory near the border of Kenya to bury US nuclear waste in exchange for a large aid program ($400 million). This completely false broadcast was quickly reported (as straight news) by newspapers in Kenya as well as by the AP and other wire services, which replayed the message as far away as Rome and Vienna.

In spite of US proof that no such meeting ever took place (the president of Somalia was out of the country at the time) and the fact that the US does not bury nuclear waste anywhere outside its own borders, the account took on a distorted life of its own. The false story, which was reprinted in numerous publications throughout Africa, produced this headline: "Africa a Global Dustbin for Nuclear Waste."

Soviet active measures operations, using forgery and disinformation, ran the gamut from clumsy and crude to sophisticated and clever. One example of the latter bore the unique distinction of being a double forgery.

The point of attack was Cyprus. The intent was twofold: to damage US–Greek Cypriot relations and to instill fear in the minds of the European public regarding US and NATO plans.

The KGB target was the Greek Cypriot newspaper *Simerini.* A document came into its possession, supposedly forwarded by a Cypriot official who, in turn, had purportedly received it from a British MP, Sir Frederic Bennett. His letter to an unnamed official was a forgery: "Your Excellency, I have received a letter which contained [sic] information about your country. Since I am not an expert on the problem of the area in which your country is located, I have decided to write to you and enclose the attached letter." The attached letter appeared on the front page of *Simerini*. It, too, was a forgery. Its author was supposedly the US Air Force vice chief of staff, who was allegedly writing to US Secretary of Defense Caspar W. Weinberger, on an urgent matter. The matter concerned the US use of bases in northern Cyprus under Turkish occupation for intended operations in the Middle East. Also, and most importantly, it concerned the evacuation of US soldiers and their families and other US citizens in case of nuclear war in Europe. The forgery elaborated on the deployment of US nuclear weapons in Europe and their possible use in support of the president's strategic schemes.

Swift action on the part of USIA P/G, USIS Nicosia, and Sir Bennett exposed the double forgery for what it was. Its publication and *Simerini's* failure to write a retraction undoubtedly had a very negative effect upon Greek Cypriots who accepted the forgery as valid, but it did not gain European press attention.

To the KGB, a forgery's failure to attain the full effect in no way impeded the continuing attack. Was the forgery's purpose to convince Latin Americans that the US was plotting with Chile to inject Chilean troops and weapons into Central America? Was the purpose to disrupt the first legitimate elections in Guatemala in 30 years by indicating US manipulation and control of the election's outcome? Was its purpose to sway the Indian public into believing that the CIA was responsible for the assassination of Prime Minister Indira Gandhi? Or was its purpose to "get away with" the murder of 269 KAL passengers by claiming that the civilian airliner was on an espionage mission?

Perhaps the forgery's purpose was all of these; in any event, the work of Section A of the KGB's first chief directorate continued without letup. Its endeavors apparently were not moderated in any form by the new spirit of Geneva or by glasnost.

Agents of Influence

Of all Soviet active measures, the most effective and the most difficult to expose was the KGB's utilization of agents of influence. Agents of influence were individuals who, either knowingly or unknowingly, sought to implant Soviet policy positions into decisions made by international organizations (such as the UN) and governmental departments. They also sought to influence congressional actions, public beliefs, and national attitudes concerning world affairs. A case in point is the former senior Norwegian government official, Arne Trehold, who frequently represented his government in negotiations with the Soviets. In June 1985, Trehold was convicted of espionage; he had been a known KGB agent of influence from 1974 to 1983.

Obviously, exposing such agents was a difficult task. At times, there were 700 Soviet diplomats assigned to West European posts. Additionally, 570 Soviets were attached to UN organizations in Europe and 1,200 Soviet bloc diplomats were serving overtly in similar capacities. Finally, and perhaps of greatest significance, there were more than 200 Soviet journalists in Western Europe. The CIA estimated that at least one-third of the nearly 2,500 Soviet and Soviet bloc diplomats and representatives (trade and commercial) were either KGB or GRU (Soviet military) agents.

An unknown percentage were engaged in political espionage, enlisting and utilizing agents of influence. Through the testimony of Soviet and bloc defectors, who were previously involved in active measures, we know that journalistic cover was a KGB priority. It gave the agent an opportunity to legitimately meet a wide range of potential contacts.

Such a person may have been an unwitting agent, a dupe of KGB subtle persuasion and psychological manipulation. Not infrequently, that person was also a journalist. Caught on the hook of his own ego by a KGB operative, who appeared to be anything but, the unwitting agent was slowly and gently played through flattery and praise. He was given a chance to contribute an article to a select group of Soviet intellectuals, after which more articles were accepted and he received payment. Inside information was exchanged for inside information until the unwitting agent began writing or broadcasting information that favored a particular Soviet line. The line could be political, military, scientific—anything that contributed to influence. In some cases the end result was blackmail; in others it was not. In all cases, however, the KGB's aim was served. There were no figures on agents of influence—only the knowledge that they were in our midst.

The term *communications warfare* sums up the totality of Soviet active measures—the use of words as weapons in all forms of media. Participants in this strategy were Soviet international fronts: propaganda organizations that conducted their activities under the semicovert direction and control of the ID. Lenin referred to the concept of fronts—groups that denied their Soviet connection while pushing the Soviet line—as transmission belts. Soviet fronts were in the vanguard of preaching, publishing, and fostering Moscow's aims.

Soviet International Fronts

The Soviets had 13 major international fronts and numerous smaller offshoots and satellites. Together they formed an interlocking network that specialized in advancing the Soviet propaganda line within noncommunist countries and the UN, and in emphasizing peace and disarmament as viewed from the Kremlin.

Months before General Secretary Gorbachev accepted President Reagan's invitation to meet in Geneva in November 1985, the fronts had received their marching orders from the ID. Meeting in Helsinki in April of that year, the leader of the 13 fronts drafted plans to launch an all-out campaign against the strategic defense initiative (SDI). The Helsinki gathering was hosted by Romesh Chandra, President of the World Peace Council (WPC, the largest and best known of the Soviet fronts). Chandra was also a member of the Politburo of the Indian Communist Party. Several weeks prior to the conference, the WPC had launched a global appeal against Washington's "space madness," proclaiming "No!" to Star Wars.

After that, the Soviet campaign to attack SDI and generate fear about it was unceasing. Through their publications, demonstrations, and forums, the fronts churned out a stream of disinformation on SDI—never pointing out, of course, that SDI is a nonnuclear system whose purpose is to destroy missiles, not people.

A broad-gauged view of the campaign occurred in Prague in July 1985 when the Christian Peace Conference (CPC) held its sixth All-Christian Peace Assembly. The CPC had been a Soviet international front since its formation in Prague in 1958; its purpose was to encourage noncommunist clergymen and their congregations to support Soviet policy, including the invasion of Afghanistan. Nearly 800 participants from 97 countries attended the assembly. At the weeklong sessions, staged to address the social and economic ills of the world, a single message overrode all else: SDI meant the militarization of space. SDI was the promotion of an aggressive policy, a web of deception. SDI was a frightening expression of human hubris.

The Soviet propaganda efforts did not succeed in changing President Reagan's strategic policy at the Geneva Summit or, nearly a year later, at Reykjavik. This lack of success, however, did not halt the ongoing Soviet effort against SDI.

Soviet international fronts never admitted defeat on any issue, and they claimed victory in President Carter's 1977 decision to forego stockpiling the neutron bomb in NATO arsenals. In following ID instructions, their emphasis could shift from the neutron bomb to INF, to chemical-biological warfare (CBW), to SDI. All Soviet policy issues played continuing roles in the fronts' propaganda.

To conceal the obviousness of their real purpose, some fronts (such as the WPC), formed additional fronts and satellites. The Vienna-based International Institute for Peace (IIP) is one example; its hierarchy was made up of high-ranking officers of the WPC. The IIP sought to attract noncommunist academics and professionals who were willing to accept the false notion that the institute was a place where scientists from East and West could frankly exchange research results on all aspects of the peace issue.

Another WPC offspring was the International Liaison Forum of Peace Forces (ILF), which was formed at WPC headquarters in Helsinki in 1973. Like the IIP, the ILF was heavily staffed by WPC officers. The ILF held a series of dialogues on peace issues; for example, opposing SDI and NATO's INF deployment. Mostly, these conferences were held in Vienna, their WPC connection carefully obscured. The ILF was turned on and off as the Soviet International Department required.

Another Soviet front that was created through WPC organizational efforts was Generals for Peace and Disarmament (GPD). Its membership comprised 13 retired senior NATO officers. Foremost among them was Italian Gen Nino Pasti, who had once served as NATO deputy commander. He and four of his fellow members were closely associated with the WPC. Unlike other Soviet fronts, the GPD held no conferences on its own. Some of its members either attended, or sent frequent statements of support to, the major front peace congresses. The members also wrote books and pamphlets that attacked NATO policies and that tacitly—sometimes vigorously— supported the Soviet line. Based in London and chaired by retired Brig Michael Harbottle, the GPD was found by the British Foreign Office to have collective views that were contrary to those of the British government and its allies. Brigadier Harbottle and some others did a considerable amount of traveling, seeking media interviews and public forums from which to speak against allied military strategy. Harbottle took credit for arranging four meetings with retired Warsaw Pact generals in which there was unanimous agreement that the United States was a threat to world peace.

The propaganda drive of the Soviet fronts for 1986 was a coordinated campaign to dominate the UN-proclaimed International Year of Peace (IYP). As nongovernmental organizations (NGO) were granted consultative status with UN economic and social departments and agencies, the fronts positioned themselves to lead the IYP celebration. (The IYP chairman, Viacheslav A. Ustinov, was under secretary general for UN political and security council affairs.)

In all, there were more than 800 NGOs, only seven of which were Soviet-directed fronts. Yet, when the 250-member Conference on Non-Governmental Organizations (CONGO) elected a 20-member board to direct operations, five of the board members were representatives of the seven Soviet fronts. In January 1986, the CONGO met again in Geneva to hold conferences under the slogan "Together for Peace." Of the

18-member executive committee chosen to organize and direct the congress, six of the delegates were representatives of Soviet international fronts—and the WPC was separately selected to coordinate CONGO activities with the UN governing body.

Throughout 1986, these particular fronts, calling themselves closely cooperating NGOs, preached the Soviet propaganda line on disarmament and peace. They collectively moved toward a climactic assembly held in October 1986 in Copenhagen under the rubric of a World Peace Congress. An attempt was made to indicate that the congress organizers had been drawn from a broad political spectrum. Such was not the case, however: The WPC—at roughly three-year intervals—was the planner, organizer, and director. The last world peace congress was held in Prague in 1983. The WPC sought to conceal its role in the Copenhagen Congress, but Denmark's Social Democratic Party (SDP) exposed it and declared that this widely heralded peace gathering was nothing more than a Soviet front operation. The Danish press reported accordingly. This exposure did not stop the event, but the SDP spokesman issued this statement: "The members of the Congress Preparatory Committee do not represent a broad selection of political affiliations. We have checked the names of the people on the organizing committee and the vast majority are Communist."

Nevertheless, some 2,000 delegates and peace activists from around the world assembled to attack US policies, support the Soviet Union, and damn SDI. It seemed appropriate that the five-day peace congress ended in a wild melee when protesters against Soviet human rights violations attempted to make their point. After that, the Soviet international fronts were less vocal and more restrained, and Soviet propaganda was promulgated more directly by Soviet entities; for example, the Soviet Friendship Society.

The Gorbachev Era

When President Reagan met with General Secretary Gorbachev in Geneva in 1985, their views of disarmament revealed major differences. There was one area, however, in which all participants at the summit appeared to be in agreement: the need to have a greater degree of cultural and personal exchange and a lesser degree of suspicion, animosity, and cold-war propaganda.

Perhaps it was a new beginning. The media proclaimed it the spirit of Geneva; Gorbachev called it glasnost (openness). It does not appear, however, that this spirit trickled down to the purveyors of Soviet active measures. Consider, for example, the following:

• The Tass disinformation report on US scientists developing ethnic weapons to be used against South African blacks was repeated as recently as June 1987.

• Post-Geneva Soviet and surrogate propaganda implied that the CIA was behind the assassination of Swedish Prime Minister Olof Palme.

• Post-Geneva Tass claimed that the US Air Force had bombed Honduran peasants with poisonous substances at night. (The charge was completely refuted by Honduran health officials.)

• Lies emanating from Moscow and carried by Soviet surrogates (e.g., the New Delhi *Patriot*) maintained that the CIA was involved in the assassination of Prime Minister Indira Gandhi, the Air India crash, and rebellion among the Sikhs.

• The Soviet press alleged that the United States continued to create biological warfare weapons when, in fact, the United States was in full compliance with the 1972 treaty that banned the production of biological and toxic weapons.

• Soviet sources alleged that the CIA perpetrated the 1978 Jonestown massacre in Guyana (a US congressman and 918 cult members were killed). The story appeared first in *Izvestiya* on 30 January 1987 and again in the Leningrad Komsomol (communist youth organization) newspaper on 10 April 1987. A particularly gruesome account, containing photographs from Jonestown, appeared in the 2–8 March issue of *Nedelya*, a Sunday supplement of *Izvestiya*. Soviet media accounts were based on the book, *The Murder of Jonestown—A CIA Crime*, published by a Soviet Ministry of Justice publishing house.

The Soviet AIDS Campaign

A recent campaign on acquired immunodeficiency syndrome (AIDS) provides insight into the methods used by the Soviets to spread a totally false story; that is, that the AIDS virus was scientifically engineered in a US laboratory at Fort Detrick, Maryland, and released into the mainstream by criminals and homosexuals who had volunteered to be guinea pigs. In October 1985, *Literaturnaya Gazeta*, attempted to launch such a campaign through the New Delhi *Patriot*, but there was only sporadic pickup by the foreign press. In May 1986, the *Litgaz* newspaper of the Soviet writer's union tried again; again, there was little outside response.

During these scattered Soviet attempts to sell the idea that AIDS was a made-in-the-USA product, US Ambassador Arthur Hartman wrote letters of protest to the editors of the publications involved, but the letters were not published. The ambassador later made the letters public, pointing out that the accusations were false and that the scientific claims were ridiculous.

The Harare Nonaligned Summit (NAM) became the launch pad for another Soviet-directed AIDS disinformation campaign that was to have far more success than the previous attempts to sell the lie. For several months prior to the NAM, Soviet and pro-Soviet media had claimed that the US government (CIA) was out to wreck the NAM, but no mention was made of the US being responsible for AIDS.

The opening attack came just prior to the summit with articles in the *Harare Sunday Mail* and the *United News of India*, reporting the distribution of a scientific tract to the delegates titled *AIDS U.S. Homemade Evil, Not Made in Africa.* (Its actual title was *AIDS–Its Nature and Its Origin*, but it sought to sell the aforementioned thesis.) The false claims of the document failed to take hold at the conference, due in part to a swift retraction by the *Harare Sunday Mail*, which carried an interview with an American virologist on the faculty of the University of Zimbabwe. The virologist found the document to be totally without scientific merit.

There was no further mention of AIDS until Novosti, the Soviet news agency, reported from Moscow that two French

164

scientists had confirmed that AIDS was the result of bacterial warfare testing in the United States. That same day in Damascus, *Al Thawra* ran the Novosti account on its front page. Two days later, "Radio Prague" broadcast the same story in its English language Afro-Asian service program. Tass then repeated the disinformation that AIDS had been developed at Pentagon laboratories at Fort Detrick, Maryland. This story was also reported in Kampala in the leftist magazine *Weekly Topic*.

Later, the London *Sunday Express*, certainly no part of the Soviet active measures apparatus, ran an interview story with Professor Jacob Segal and his wife, Lilli, principal authors of the Harare Summit AIDS document. The *Express* reported that Segal, and to a lesser degree his specialist colleagues (Dr John Seale of London and Dr Robert Strecker of Glendale), contended that the killer AIDS virus was artificially created by American scientists during laboratory experiments which went disastrously wrong and that a massive cover-up has kept the secret from the world until today. This totally erroneous front-page claim had the effect the Soviets sought. Coming from London in a conservative publication, the disinformation caught hold.

In the week that followed, 14 USIS posts reported reprintings of the *Express* story. By end of the month, the total approached 30. Publications in Africa, Latin America, the Middle East, India, and Western Europe carried the disinformation. In Wellington, New Zealand, The *Dominion* ran the story on its front page under the heading, *Bizarre Theory on AIDS Origin*. In Helsinki, although a tabloid carried the piece, it also carried this quote from a noted Finnish AIDS expert, Dr Jukka Suni: "I know of Doctor Segal and of his reputation. He is a 'prophet of doom' who has been getting worse year after year."

In Eastern Europe, only in Poland was the *Express* article replayed. It was featured in the Polish government daily, which made a connection with the *Literaturnaya Gazeta* account of October 1985. Several days later, the Soviet press weighed in through Tass and *Pravda*. The Soviet official news agency used reports from its New York correspondent, who had cited out of context a National Academy of Science report on AIDS. Tass attempted to link factual scientific information with Segal's

Fort Detrick disinformation. Segal was again referred to inaccurately as a French scientist.

Pravda went a vicious step further, running an editorial cartoon that showed a man in medical dress handing a beaker full of swastikas suspended in liquid to a second man wearing a US military officer's uniform. The officer is handing the doctor a wad of dollars. The beaker is labeled "AIDS virus." A number of corpses lie on the floor. Ambassador Hartman again wrote letters of protest, this time to the editors of Tass and *Pravda*. As before, he received no response.

Result of USIA/State Response

Although the disinformation campaign was initially successful, rebuttals from USIA and State Department INR quickly helped to reverse original media acceptance of the Soviet effort. Adding to this was the stabilizing effect of two widely attended AIDS conferences in Brazzaville and Berlin. In Brazzaville, where representatives of 37 countries gathered under the chairmanship of Dr Johnathan Mann, AIDS director of the World Health Organization, no mention was made of the campaign. In Berlin, it was dismissed as being unworthy of comment.

As a result, many of the publications that had carried the *Sunday Express* story printed retractions (which included a modified backdown by the *Express* itself). In a stinging editorial, the *Hindustan Times* declared that not a single reputable scientific journal had supported the thesis that AIDS was man-made. Almost every Western specialist is convinced that the AIDS virus mutated naturally and spontaneously from an animal virus.

In Brazil, *O Estado de S. Paulo* observed that AIDS and *disinformatza* are very similar in their corrosive action: One infects and destroys the body while the other succeeds in shaking and disorienting the soul. There are two basic questions at the root of this Soviet active measures campaign: Where did the claim originate? Who are the scientists who have attempted to spread them?

In July 1983, the New Delhi *Patriot* published a letter from an unnamed "well-known American scientist and anthropologist who wished to remain anonymous." The letter was a lengthy undocumented account of how the AIDS virus was developed by the CIA and the Pentagon. Fort Detrick, Maryland, was reported by the writer to be the place where the disease was created. He added just as incorrectly that a similar US laboratory in Lahore had bred super mosquitoes and other insects that could spread dangerous diseases such as yellow fever, dengue, and encephalitis. The only source the unknown writer gave for any of this information was the totally unreliable cult organization, the Church of Scientology.

Literaturnaya Gazeta used this rambling piece of gobbledygook as its source in its October 1985 story. So did the Segals in their document presented at Harare. Jacob Segal was born in Leningrad and was a longtime resident of the Soviet Union before moving to East Berlin. He is a 75-year-old retired biologist and self-announced AIDS specialist who failed to attend the mid-November Berlin conference on AIDS. He maintained that AIDS was first tied to Fort Detrick by a British researcher who published an article in a New Zealand journal. Segal's wife, Lilli, however, offered a different source—the East German Urania Press. Urania's prime function was to disseminate Soviet propaganda.

Lilli Segal is a retired researcher and professor of epidemiology. She survived Auschwitz and has been a resident of the GDR since 1953. She and her husband worked as biology instructors in Cuba for several years. Both have been retired from Humboldt University since 1983. Their AIDS thesis was absurd, according to world reknowned AIDS specialists. It had no bearing on serious scientific knowledge of the virus. It was on a par with the Segals' attempt to link the surfacing of the disease in New York City with its proximity to Fort Detrick. Their report stated that criminals who had engaged in homosexual practices during the long time of their imprisonment "obviously" concentrated in the nearest big city after their release and it was therefore logical that the first AIDS cases should have been registered in New York. In fact, however, New York is 250 miles from Fort Detrick while Baltimore is only 45 miles and Washington, D.C., 50 miles from that installation. Further, President Richard M. Nixon called a halt to all offensive biological warfare experimentation in 1969. In its place, a vaccine program was instituted at the US Army Medical Research Institute of Infectious Diseases. The laboratory's work was focused on finding vaccines *against* biological warfare.

Fort Detrick is not used entirely for military purposes. Of the 25 tenants on the installation, one is the Frederick Cancer Research Facility. It operates under the direction of Health and Human Services with a department devoted to AIDS research. It has never been a military component. Other research is conducted by the Department of Agriculture in both foreign and domestic crop diseases. Fort Detrick's raison d'être is to cure, not to inflict; to prevent, not to infect.

It should be added that neither Dr John Seale, London venereologist, nor Dr Robert Strecker, California gastro-enterologist, both of whom were quoted in the *Sunday Express*, subscribed to the disinformation regarding Fort Detrick.

Conclusion

The twisted course of the Soviet active measures AIDs campaign can be traced back to a Soviet-aligned Indian newspaper. It quoted an unnamed individual whose line of lies took root in the Soviet press and briefly flourished via a Soviet-bloc document that was completely lacking in scientific credibility. Later, a new piece of AIDS disinformation emerged from India. It attempted to link the disease to a Union Carbide pesticide. Still later, a Communist party member in the Indian parliament sought to press the AIDS-made-in-the-US line. Nevertheless, the disinformation boomeranged on the Soviets; many publications around the globe have come to recognize and understand the nature of this Soviet active measure.

Appendix A

Weinberger Forgery

A few years ago, a forged document purporting to be a speech by US Secretary of Defense Caspar W. Weinberger surfaced in the Federal Republic of Germany. The document is dated 25 November 1983. No such speech was ever made, and Secretary Weinberger never made the remarks attributed to him. The document was a complete fabrication. Its aim was to convince Europeans that the US purpose in developing the Strategic Defense Initiative (SDI) was not only to gain US military superiority over the USSR but also to dominate its NATO allies. Both goals were supposedly to be accomplished largely by US control of space through SDI, which was described in the forgery as an offensive system.

The concepts used in the document did not reflect US government thinking, but were, in fact, what the Soviets wanted West Europeans to believe about US views. Although for the most part the forgery was written in acceptable English, several mistakes in word usage indicated that the drafter's first language was not English. For example, it asserted that the United States sought "prevalence" over the Soviet Union at all (military) levels. The forger(s) confused *prevalence* with *superiority.*

A major point made in the forgery was that the United States would suffer a disastrous depression without SDI. However, the key and distinctive error had the secretary of defense commenting that the Soviets have not been engaged in an SDI program of their own. In fact, Secretary Weinberger, the president, and other leading US defense spokesmen had pointed out repeatedly that Soviet scientists had been at work on a similar project since 1968.

The Weinberger forgery is one of several addressed in this report. All sought to undermine US policies and intentions.

Disinformation Hits Home

Soviet disinformation attempted to "hit home" with a forged letter purportedly written by Herbert Romerstein, United States Information Agency (USIA) Soviet active measures coordinator to US news organizations. The *Washington Post* approached USIA with the forged letter, which Mr Romerstein had supposedly sent to Senator-Select David Durenberger (R.-Minn.), chairman of the Senate Select Committee on Intelligence. The forgery was cleverly conceived but poorly executed. Its purpose was to smear USIA by making it appear that the agency was proposing that the United States should make false claims about the Chernobyl nuclear accident.

The forgery alleged that Mr Romerstein proposed to Senator Durenberger that USIA disseminate rumors about the accident. The forgery suggested that "our associates in European information media" spread reports that would include the following details:

- the "number of victims should be alleged to be somewhere between 2,000 and 3,000";
- a "mass evacuation of population from the 100-mile zone" around Chernobyl should be alleged to have occurred; and
- "transport problems, shortage of various goods, chaos, and panic should also be given publicity."

The forgery also recommends that "our allies should be influenced so as to make a request for compensation for contamination of their territory." The document takes advantage of the insights offered by hindsight. It is backdated to 29 April 1986, the day after the Chernobyl accident was first reported by the Soviets and just before rumors (of exactly this nature) began to appear.

Had the forgery been accepted by the press, it would have given currency to the notion that rumors about Chernobyl had been generated by USIA rather than by the Soviets' refusal to release sufficient data about the events surrounding Chernobyl. At the same time, USIA would have been cast as a manipulative, rumor-mongering propaganda apparatus free to dictate to senators and unnamed "associates" in the European media.

Poor execution, however, ruined the Soviet attempt. The forgery was not composed correctly and it contained an incorrect address for Durenberger. Its falsity was easily ascertained by journalists who telephoned the USIA and Senator Durenberger's office.

Ironically, Romerstein's forged signature was taken from a letter he had previously sent to Lt Gen Robert L. Schweitzer, head of the Inter-American Defense Board, informing General Schweitzer about a forgery purportedly written by him. Even more obvious, the word *copy* in Romerstein's handwriting at the top of the Chernobyl forgery is identical to a notation made on copies of the letter to General Schweitzer.

The forgery had all the hallmarks of a Soviet active measure, although its sloppiness probably indicated an operation executed by personnel in the United States rather than those in the Moscow Center, who could be expected to produce a more professional product. In analyzing forgeries, one key is to determine who benefits. Certainly, Soviet purposes would have been served if the USIA had been blamed for the bad press that was actually caused by Soviet intransigence and stone-walling over Chernobyl.

Levchenko on Dobrynin

At a USIA briefing on 20 August 1986, at Washington's Foreign Press Center, former KGB major and active measures specialist Stanislav Levchenko made some observations about Anatolii Fedorovich Dobrynin, former Soviet Ambassador to the US who later headed the Soviet Communist Party Central Committee's International Department.

Mr Dobrynin's appointment to the International Department post had been the subject of much speculation in the American press. As a former staff member of the International Department, Levchenko's perspective on this issue was instructive.

Basically, Levchenko discounted the theory that Dobrynin's appointment was "positive for the future development of healthy relations between the Soviet Union and the United States."

> I dare to disagree with this point because if Mr Gorbachev really wanted him to do that kind of job, he could have made him, let's say, first deputy foreign minister in charge of bilateral relations with the United States, and he could spend all his days and nights working in that field. Instead, Anatolii Dobrynin becomes chief of the International Department, which, in another parlance, is the largest subversive mechanism in the world. The purpose of that department is not enhancement or deepening of bilateral or multilateral relations at all. It's just the contrary. It is the department which, among many other

functions, is disseminating disinformation in the interests of the politburo and running all sorts of operations in that field.

Levchenko believed that Dobrynin would reinvigorate the active measures operations of the International Department and use his knowledge of the United States to bring a new professionalism to its activities.

> He, of course, knows about some flaws which the active measures machine had during the last couple of years when his predecessor, Boris Ponomarev, was really getting old. I personally have no doubt that Mr Dobrynin will try, within a rather short period of time, probably within this year, to restructure the mechanism in such a way that it will work better, on a much more professional basis, that sloppiness will probably disappear, and that the scale of operations against primarily the United States and NATO countries on a global basis will definitely be enhanced. That's my projection.

Levchenko's final assessment on Dobrynin: "He is the person who has close to perfect knowledge of the United States. He's probably one of the best specialists in the Soviet Union on the United States, who is generally considered to be a very talented, bright person who is, of course, a devoted Communist and quite a devoted person to the cause of the politburo."

National Security Council Forgery

A very well-done and ambitious forgery surfaced in Lagos, Nigeria. It bears the title, "US Strategy in Foreign Policy 1985–1988" and falsely claims to be a "copy of a summary paper on US foreign policy approved by the (United States) National Security Council (NSC) in February 1985." A section of the forgery dealing with purported US policy in Africa was published in Lagos in the weekly magazine *Africa Guardian*. Its broad scope and relative professionalism increase the chances that it may reappear in other areas.

The forgery was a 10-page, single-spaced document lacking any identifying symbols and classification markings. It was divided into eight topical sections dealing with US policy toward important regions. It was cleverly crafted, skillfully weaving together US policy with phony assertions that would outrage foreign as well as domestic audiences. For example, the United States was falsely portrayed as believing that "a preemptive strike potential, together with guidance, communications and intelligence systems of high sophistication, would guarantee our security to a degree which would permit us to exert severe pressure on the Soviet Union to the point of issuing an ultimatum if necessary."

Other false assertions were that:

- The United States was seeking "the establishment of an effective first-strike (nuclear) capability by the year 1995."

- "Strategic superiority" was the goal of US policy, and advances in military technology might "suggest" or "even demand a preemptive strike."

- NATO allies should be pressured "to conform to US policies."

- The US problem with the Greek leadership was that it had an "independent attitude."

- The United States is opposed to curbs on nuclear and chemical weapons.

- In Nicaragua, "US military intervention cannot be ruled out, keeping in mind the lesson of Grenada."

- The Organization of American States "should be made to accord with our national interests."

- A negotiated settlement between Vietnam and an ASEAN bloc over Cambodia "in no way corresponds with US interests."

- Japan must agree to a stronger US military presence in that country and to increase its military strength through the purchase of US arms.

- The United States will "seek to hold China to policies corresponding with US global interests."

- The United States feels that China has an "unpredictable political future" and an "absence of stable power succession procedures."

- The United States opposes the successful implementation of China's modernization program because it "might lead to rivalry or even confrontation with the United States."

- "Nothing short of Qadhafi's assassination will bring any significant change in Libyan policies of support for international terrorism."

- Israel is referred to as "the Jerusalem government," implying recognition of Jerusalem as the capital.

- The United States is planning to use the rapid deployment force for "preemptive operations" if a "threat to our interests" arises.
- The Soviet presence in Afghanistan is seen as a "propaganda bonus" which can be utilized to increase US military and political power in the Persian Gulf region, perhaps even leading to "the restoration of American influence in Iran."
- The main lines of US policy in Africa are to exploit the area as a necessary source of raw materials and to control African affairs in conjunction with South Africa.

Many other themes and subthemes were planted in this lengthy forgery. As with the other forgeries mentioned above, this one appears to be a product of the Soviet Union or one of its surrogates. The themes cited reflect Soviet claims about US policy.

The Assassinations of
Olof Palme and Indira Gandhi

If one were asked to name the major similarity between the gunning down of Swedish Prime Minister Olof Palme and the gunning down of Indian Prime Minister Indira Gandhi, the answer would be that the Soviet Union, by innuendo, falsely accused the Central Intelligence Agency (CIA) of committing both crimes and repeated the lie through its propaganda outlets.

On 31 October 1984, Prime Minister Indira Gandhi of India was assassinated by Sikh members of her bodyguard. Several hours after the murder, "Radio Moscow" declared the assassins had "received their ideological inspirations" from the CIA. "Radio Moscow" broadcast a familiar litany of disinformation, accusing the CIA of guilt in the assassination of world leaders over a 25-year period.

Pro-Soviet Indian publications were quick to pick up the Moscow line. They added their own embellishments, some directly accusing the United States of having murdered Mrs Ghandi. These claims were replayed by the Soviet press. Responsible Indian newspapers would have none of it, however. A *Hindustan Times* editorial summed up public reaction: "The Soviet media saw in the grim event a golden opportunity to mount cold war propaganda against the United States. . . . Cold War propaganda on such a tragic occasion by the Soviet Union is uncalled for and most unfortunate." President Reagan called the Soviet allegation, "the world's biggest cheap shot in a long, long time."

The Soviet lie became a double lie when Secretary of State Shultz confronted Soviet Prime Minister Tikhanov at Mrs Ghandi's funeral and Tikhanov denied that his country was making such claims. The Indian government, in a determined investigation of the assassination, found no trace of any US involvement.

These events occurred about a year before President Reagan met with General Secretary Gorbachev in Geneva and some 15 months before the murder of Swedish Prime Minister Palme. It was hoped that the Geneva meeting would bring a change in the kind of outrageous disinformation that had been spewing out of Moscow day in and day out.

Prime Minister Palme was killed on a Friday night. The next day, Tass published a commentary on Palme's death. The commentary was written by Anatoly Krasikov, deputy director of Tass. Krasikov's commentary, according to the Swedish daily *Aftenpost,* "indicated that CIA is behind the murder." This commentary employed exactly the same kind of slippery innuendo "Radio Moscow" had used in implying that the CIA was responsible for Indira Ghandi's assassination.

The next day (Sunday, 2 March), *Pravda,* the official Communist Party newspaper, added its commentary. The writer, Chingiz Aytmatov, academician of the Soviet Academy of Sciences and delegate to the 27th Congress of the CPSU, raised a basic question and then answered it. Who was, or who were, the assassin(s)? We do not yet know, but it is clear even now that the terrorist crime was committed by forces that are interested in shifting whole countries and regions to the right and in intimidating the public, which heeds the voice of an honest individual.

Response from the Swedish newspaper *Morgenbladet* was swift. In an editorial titled "Grotesque," it attacked Gorbachev's "mouthpiece" in choosing to exploit the situation at a "tragic moment." It went on to say, "the real question is what Gorbachev means with his fine words about détente and peace when he allows such gross propaganda. There is brutal contradiction between the hopes he has expressed for a relaxation of internal tension and his continued hammering on public opinion with anti-American propaganda."

On Monday, the New Delhi Hindi daily *Navbharat* carried a Moscow dateline report: "The official agency Tass has alleged that the assassination of Swedish Prime Minister Olof Palme was the handiwork of the CIA." Other Indian publications, particularly those that were pro-Soviet, repeated the accusation and parallels were drawn between the political viewpoints of Mrs Ghandi and Mr Palme.

In the course of the next few days, the Tass-*Pravda* effort was replayed in a number of locales—Ghana, Malta, Australia, Uruguay, Peru. In the latter two countries, the publications were Communist Party weeklies. In Australia, the conservative daily *Australian* concluded an account by its Moscow correspondent with a quote from former Swedish Prime Minister Falldin: "I refuse to believe it is a political assassination. . . . This must be the work of a lunatic."

Elsewhere around the globe, the Soviet accusation was ignored; but that was not the end of it. Seventeen days after the murder, Aytmatov, the writer of the *Pravda* commentary, was interviewed by the Swedish newspaper *Dagens Nyheter*. "Aytmatov was now clearly alarmed that his words would be interpreted" to suggest the CIA had committed the crime. But the *Pravda* writer's retraction was something less than unequivocal: "The people who murdered Olof Palme do not need to come from the United States or the CIA—even though it does seem to be the case of reactionary forces."

The Soviet academician was simply following the standard Leninist technique of hedging his *Pravda* commentary with obfuscation. However, just as the Soviet-oriented accusations of CIA guilt in the assassination of Indira Ghandi continued to surface, so, too, did those about Olof Palme. His name was added to the long list of assassinated leaders that Soviet disinformation organs recited on a regular basis as victims of CIA operations.

Yellow Rain Again

Certain characteristics in Soviet active measures operations followed a consistent pattern. One was to try and reverse blame for Soviet Military actions by accusing the United States and its allies of carrying out exactly the same acts. This was done not only by describing the Soviet techniques originally followed, but also by employing the same terminology; for example, *yellow rain.*

In March and November 1982, the US State Department presented to the United Nations reports of "Soviet involvement in provision and use of toxin weapons." The reports offered "information, evidence and an analysis of results the United States had obtained on the use of toxin and other chemical war agents by the Soviets in Afghanistan, and by the Lao and Vietnamese under Soviet supervision in Laos and Kampuchea." The victims referred to chemical agents that had been sprayed and dropped on their villages by plane, helicopter, and artillery shell as yellow rain. Between August 1983 and February 1984, further evidence of yellow rain attacks in Southeast Asia were submitted by the United States to the UN. These charges were angrily denied by the Soviets, who submitted their own scientific paper. It maintained that Agent Orange, the defoliant used by the US to clear jungle areas during the Vietnam War, had contaminated the elephant grass. Its infected spores, the Soviet government claimed, had been carried by prevailing winds into Laos and Kampuchea where they had tainted the environment and brought illness and death to those affected. Scientifically, this explanation was recognized as hogwash.

By the end of 1985, reports of yellow rain in the afore-mentioned areas had "diminished in number." But on 7 February 1986, yellow rain was suddenly in the news again. This time, the area was Honduras and the culprit was the US Army. The newspaper *Tiempo,* reporting from San Pedro Sula, quoted a leftist trade union publication which maintained that the US Army was spraying lethal chemicals on civilians in northern municipalities for experimental purposes. The chemicals, it reported, caused fainting and convulsions. Further, it said that strange sores appeared on the bodies of children. The report said that the affected communities had requested investigations by health authorities.

In the first published accounts of the reported problem, no mention was made of yellow rain. On 22 February, Tass quoted *Red Star,* the official Soviet Ministry of Defense newspaper: "The United States is carrying out criminal experiments with chemical weapons in Honduras. . . . U.S. aircraft have been scattering yellow toxin in the air similar in its characteristics to the highly toxin 'Agent Orange.'. . . Eyewitnesses to the secret flights by the US aircraft say that 'yellow rain' falls every time the aircraft appear."

That same day, "Radio Moscow" offered a commentary, repeating the same accusations and using the same terms. But the accusation went even further. The commentator said the Pentagon was preparing to launch chemical and biological warfare against Nicaragua and was responsible for an epidemic of dengue that had hit that country earlier. Some of these charges were picked up by the Panamanian press, which added that the yellow rain story had been born of a theory "between groups of the Honduran Left."

Previously, the US Embassy in Tegucigalpa had denied that US forces were testing chemical weapons anywhere in Honduras, "least of all against the civilian population." Not only had a US medical team investigated the charges, but so had the US military and a Honduran congressional group. Under the heading "Congressmen find downpour of calamities instead of yellow rain," the newspaper *La Prensa,* along with other Honduran publications, including *Tiempo* (which had broken the original story), cited the cause of the problem as an epidemic of mange stimulated by a lack of hygiene; the germ had spread through physical contact.

The daily *El Haraldo* carried an interview with Honduran Army Chief of Staff General Humberto Regalado Hernandez, who said that a military investigation coincided with state and private reports. The disease, he said, had been known since biblical times as scabies and was not the result of chemical gases dropped from unidentified aircraft on evening flights. The only flying in the area had been done by Honduran crop dusters who were spraying sugar cane.

The general added that he believed the purpose of the stories was to discredit US military detachments that were training in Honduras. Another possibility, he said, "is that the USSR, via Cuba and Nicaragua, is using sophisticated methods. . . . The presence of 'yellow rain' in our homeland was ruled out, but there is evidence that the local fifth column is carrying out disinformation campaigns ordered by the Soviet connection in Tegucigalpa." This Soviet yellow rain story was turned off by the combined actions of the Honduran government and the press.

Appendix B

Former Soviet International Front Organizations

Afro-Asian People's Solidarity Organization (AAPSO)

Christian Peace Conference (CPC)

International Association of Democratic Lawyers (IADL)

International Federation of Resistance Fighters (IFIR)

International Institute for Peace (IIP)

International Organization of Journalists (IOJ)

International Radio and Television Organization (OIRT)

International Union of Students (IUS)

Women's International Democratic Federation (WIDF)

World Federation of Democratic Youth (WFDY)

World Federation of Scientific Workers (WFSW)

World Federation of Trade Unions (WFTU)

World Peace Council (WPC)

Appendix C

Generals for Peace and Disarmament

British Guiana
Generalmajor Gert Bastain

Canada
Major General Leonard V. Johnson

Federal Republic of Germany (West Germany)
Generalmajor Gunter Vollmer

France
Admiral Antoine Sanguinetti

Greece
General Georgios Koumanakos
Admiral Miltiades Papathanassiou
Brigadier Michalis Tombopoulos

Italy
General Nino Pasti

Netherlands
General-major Michiel H. von Meyenfeldt, Chairman

Norway
General Johan Christie

Portugal
Marshal Francisco da Costa Gomes
General Rangel de Lima

United Kingdom
Brigadier Michael Harbottle OBE

What Soviet PSYOP Mirror Imaging Can Tell Us

James Melnich

Soviet psyoperators were sometimes ignorant of significant nuances concerning the societies they targeted for propaganda development. Though time and experience tended to improve certain technical aspects of Soviet PSYOP, Soviet psyoperators sometimes continued to reach down into their own limited Soviet experiences for examples and themes that they projected onto the Western societies they were attacking in their propaganda. Study of this mirror-imaging process can be very revealing and may point out important weaknesses. This is true even though some segments of the Soviet PSYOP hierarchy appeared to have gained in sophistication. Identifying some of these tell-tale characteristics will help us to assess the overall Soviet PSYOP threat.

Glasnost, Gorbachev, and Disinformation

The campaign of glasnost in the Soviet Union under Mikhail S. Gorbachev, with its emphasis on a more open Soviet press, by no means signaled an end to Soviet disinformation. On the contrary, since the press in the USSR began writing on many topics previously considered taboo, there were more opportunities for mirror-imaging disinformation to present itself—perhaps with some new twists.

For example, we can examine one of Gorbachev's perceptual blunders concerning the United States and see how it relates to the concept of mirror imaging. In early 1987, Gorbachev met with several US congressmen in Moscow. In the course of their discussions, he suggested that America create separate states for blacks, Puerto Ricans, and citizens of Polish background.[1] Apart from this being a ludicrous and insulting

191

recommendation, it provides some insights into the Soviet system. To begin with, it demonstrates Gorbachev's ignorance of the internal policies of the United States. Further, it assumes that we have a Polish problem that must be dealt with. From the standpoint of mirror imaging, it suggests that the Soviets dealt with many of their minority problems by sending minorities to autonomous regions somewhere in the Soviet Union.

The Soviets had serious minority problems, especially with their Central Asian Muslims. There have been riots in Kazakhstan, demonstrations in the Baltic states, demands by Crimean Tatars for their rights, and pressures for greater Jewish emigration. Demographic analysis of the Soviet population suggests that the Great Russian nationality itself was becoming a minority in its own country as Central Asian birthrates skyrocketed.

As Soviet policymakers sought answers to these problems, Soviet propagandists were producing disinformation about an alleged Western ethnic bomb supposedly targeted against US minorities. This propaganda is an excellent example of mirror imaging. What some in the Soviet Union undoubtedly wished they themselves had, they projected as a policy practiced by their ideological enemy. This disinformation example shows us both the Soviet propagandists' intense prejudice and how limited their worldview really was. It is an example that is repeated over and over in a variety of contexts, whether one is considering AIDS, baby parts, or chemical weapons.

One might add that, although the Soviet Union did have its share of extremely sophisticated propagandists, such as Georgii Arbatov of the Institute of the USA and Canada, Politburo member Aleksandr Yakovlev, and Vladimir Posner of Radio Moscow, the fact that Gorbachev himself could make such an error should tell us something about the attitudes of rank-and-file Soviet propagandists.

Ultimate Enemy Projections

Soviet propaganda was rife with allusions to the ultimate enemy, whether that enemy be world imperialism, Nazi storm troopers, supporters of SDI, Zionism, or religious activists in the Soviet Union. The ultimate enemy was usually portrayed as being antihuman and opposed to everything the Soviet Union stood for. For example, Zionism was defined as a "deadly enemy of the Soviet State from its very beginning."[2] There was usually a plurality of Soviet ultimate enemies, depending on the international situation. An ultimate enemy in Soviet propaganda was first dehumanized, then made part of an anti-Soviet worldwide conspiratorial network (i.e., Afghan freedom fighters became bandits fighting as agents of US imperialism; unofficial religious groups in the Soviet Union were linked with dark forces from abroad).

In attempting to describe the ultimate enemy, whoever and whatever that might have been at a given time, Soviet propagandists sometimes drew from actual Soviet acts of barbarism. Thus their disinformation, as a form of mirror imaging, reflected back into the pool of Soviet reality. For one example, I will draw from personal experience.

In the early 1980s, I met a Soviet émigré woman who was convinced that evangelical Christians in the Ukraine sacrificed babies by rolling them in barrels with spikes as part of some religious ritual. She, of course, had never witnessed such a thing, but said that she had heard about those people when she lived in the Soviet Union. Obviously, she was the victim of very gross Soviet disinformation—and there were other Soviet campaigns in which other religious groups were under attack by the state. One can find numerous articles accusing Soviet Baptists of drowning children or performing other dark and perfidious acts—the whole purpose being to defame the targeted group and further isolate them from the general population.

Nevertheless, the matter of the spikes continued to bother me: why such a particular disinformation image? Where did it come from? Did a propagandist simply make it up out of thin air, or did it spring from some other source? I have no final answer to this question, but a year or so later, when I was reading Michael Voslensky's work, *Nomenklatura: The Soviet Ruling Class*, a particular passage leapt out at me. Voslensky, in quoting a 1920s account about the Cheka (the earliest forerunner of the KGB), recounted various Cheka methods of torture and execution of their victims. One such method was reported this way: "at Voronezh they put their victims naked in barrels spiked on the inside and rolled them."[3] This account and recent Soviet disinformation against religious groups are separated by more than one-half century. Are they somehow related, or did later Soviet propagandists simply make up the recent vicious accusation out of thin air? One cannot be certain, of course, but I would posit this as a possible example of a mirror image projected out of the past.

Zionism and Nazism

In examining the concept of ultimate enemy in Soviet propaganda, we might also consider Soviet disinformation that compared and linked Zionists with Nazis. One would be hard-pressed to find a more vicious lie, but such comparisons fit Soviet concepts of the ultimate enemy very well. They also furthered Soviet foreign policy goals in the Middle East. Virulent anti-Semitism was imbedded in Soviet society, along with fears of a Jewish conspiracy dating back to the fraudulent Protocols of the Elders of Zion. There was also the memory of Nazi fascism, which cost 20 million Soviet lives during World War II and the fear that German revanchism would once again rear its ugly head. Soviet propagandists simply combined the two images.

What this tells us is that the Soviet propaganda machine (regardless of its short-term goals for the propaganda, such as defaming Israel in third world countries) was itself convinced that there were many real enemies, both within and outside the Soviet Union.[4] These were then reduced to extremely crude and violent images and made a part of an ongoing international conspiracy. They probably also believed that such ultimate enemy projections were both necessary and useful for their own population. While many societies resort to ultimate enemy projections and stereotypes in propaganda during wartime, the Soviet Union appears to have been one of the few nations to do it consistently over time during periods of peace.

As a footnote, one might consider how sensitive the Soviets themselves were to being stereotyped in the West. The TV series "Amerika" showed the Soviets as adversaries, but adversaries possessing many human characteristics—a far cry

194

from the ultimate enemy. Yet the program was denounced in almost apocalyptic terms by Soviet spokesmen—and threats were made prior to its airing. Soviet propagandists continued to show daily hatred against the West on a massive worldwide scale, with only an occasional protest from the West when the results of a given lie would have had serious and undeniable consequences.

The Invasion of Grenada

In late 1983, the Soviet newspaper *Izvestiya* not only attacked the United States for invading Grenada but also accused US forces (allegedly quoting the Mexican newspaper *El Dia*) of using chemical weapons to poison some 2,000 Grenadens, including women and children, and of recording their suffering and deaths on film. This gruesome fabrication, which was read by millions of Soviet citizens, further stated that the bodies were shipped back to the United States for additional study. The author of that article was A. Kuvshinnikov.[5]

For a long time I tried to discover who A. Kuvshinnikov was or is, or whether it was a pseudonym. Then another article by A. Kuvshinnikov appeared in *Izvestiya* on 21 August 1987. This article was said to be from the US correspondent at the USSR Foreign Ministry Press Center.[6] It is likely that these articles were written by the same person. In his especially pernicious article about Grenada, Kuvshinnikov attempted to set up a parallel between the WWII Nazi death camps with their human experiments and the US invasion of Grenada. This is another example of the ultimate-enemy image. No citation was given for *El Dia*, which was described by *Izvestiya* as an influential Mexican newspaper.

Soviet propagandists apparently believed that their audience would accept such base lies to some extent. And were these articles also designed to mitigate the effect of certain Soviet actions? In my opinion, they may tell us in part what the Soviets themselves have done or are capable of doing. The horrors of Afghanistan, as recounted by refugee survivors of Soviet atrocities, come to mind. Were the consciences of some Soviet veterans returning from Afghanistan (who may have participated in brutal crimes against civilians) eased when they thought the United States had done worse?

Mirror imaging in Soviet propaganda is, I believe, an analysis area of great value to all who study the former Soviet Union—especially those who seek to understand the nature and depth of Soviet PSYOP.

Notes

1. *Novoe Russkoe Slovo* (The New Russian Word), 19 April 1987.

2. See V. Alekseev and V. Ivanov, "Zionism in the Service of Imperialism," *International Affairs* 6 (1970): 59, in Baruch A. Hazan, *Soviet Propaganda: A Case Study of the Middle East Conflict* (New Brunswick, N.J.: Transaction Books, Keter, 1976), 150.

3. Michael Voslensky, *Nomenklatura: The Soviet Ruling Class* (New York: Doubleday & Co., 1984), 279, in P. Milyukov, *Rossiia Na Perelome* 1 (Paris, 1927): 193.

4. For a fascinating account of alleged Politburo views of worldwide conspiracies, see Arkady N. Shevchenko, *Breaking with Moscow* (New York: Alfred A. Knopf, 1985).

5. A. Kuvshinnikov, *Izvestiya*, 9 December 1983, 4.

6. Ibid., 21 August 1987, 4.

Indoctrination of Hate

Lev Yudovich

Soviet military doctrine required the soldier to develop a strong feeling of hatred. This hatred was cultivated so the soldier would (1) believe in victory, (2) be willing to act in a dangerous situation, and (3) be psychologically prepared to operate under the conditions of modern war.

> In case of war, the Soviet soldier will face a strong and brutal enemy who has been well indoctrinated in a spirit of irreconcilability toward the Soviet people. The Soviet soldier must therefore have a clearly defined attitude for the enemy: hate, contempt, and a feeling of superiority over them. We must show the soldier the strong and weak points of our enemies. We have to define the enemy in terms of his psychological characteristics (or stereotypes) that are related to his nationality or ethnic group.[1]

The Soviet soldier was constantly being indoctrinated to develop hatred for the West. The indoctrination of hate started with the myth of a US threat to the Soviet Union. This myth allowed Soviet political and military leaders to keep the entire country in a state of tension and stress. The myth also made it easier to justify military spending and the consequent lowering of standards in domestic life. The American threat to the Soviet Union was the fundamental thesis of Soviet propaganda.

The aim of Soviet propaganda was to create a stereotyped pattern of thinking and reacting. The myth helped Soviet military and political leadership instill prejudice in the soldier. He was trained to become suspicious of foreigners and their way of life. His suspicion and deep distrust became the basis for hate. According to the Soviet view, "the political myth is an

This article provided courtesy of the US John F. Kennedy Special Warfare Center and School, Fort Bragg, North Carolina. Part of an original work titled, *An Assessment of the Vulnerability of the Soviet Soldier.*

ideological tool of the imperialists and a tool for psychological manipulation of the masses" (Col Gen D. Volkogonov). Thus the Soviet authorities use the very same technique that they accuse the enemy of employing.

The Soviets instilled hatred for the United States even in peacetime. They justified this approach by explaining that, in the past, the final moral hardening of the soldier could be achieved during the course of combat. In the future war, however, there will not be enough time for psychological preparedness. The soldier must, therefore, be prepared through training. He must put aside any moral convictions he might have and be oriented only toward accomplishing his mission.

Hate produces excitement—and excitement can overcome the soldier's fear. To develop hatred, political workers used a wide variety of indoctrination methods (e.g., discussions in small groups, lectures, and political reviews) and every form of media. Deception, falsification, discreditation of Western leaders, and even cartoons became methods for building up hate. The US soldier was shown as a hireling, a mercenary, and a freelance murderer with a very low moral level. He had no convictions, and his main reason for fighting was to get money. The Soviets stated that the US soldier was not reliable in Vietnam. As the military situation in Vietnam got worse, said the Soviets, there were more deserters, and racial tensions increased. The reason they presented such distorted views was simple—to cultivate hate.

In 1952 the Military Publishing House of the Ministry of Defense of the USSR published a book written by Col Polkovnik B. W. Karpovich titled *The Ideological Indoctrination of Soldiers in the Armies of the United States and Great Britain.* This book was aimed at military readers. The author pictured American soldiers as fascists and animals.[2] When a young man enters the US Army, he stated, all training is aimed at turning him into a skilled murderer.[3] American officers encourage US soldiers to be murderers and rapists. In the Korean War, according to the colonel, all American soldiers raped Korean women and girls, and officers created special brothels in which Korean women were forced against their will to provide sexual services for the soldiers.[4] American soldiers, he said, far exceed the Nazis in their gangsterism, brutality, and greed.[5]

Although those statements were made around 1952, this method of describing American soldiers persisted. Despite détente, indoctrination of hate continued. In 1986, for example, the Main Political Administration ordered the use of a film, "Hate the Imperialists,"[6] and a 1968 article, "The Psychology of Murder and Torture," which purported to picture the American soldier.[7] Political workers indoctrinated hate during every major maneuver from Dnepr to Shield '84.[8] One dramatic example of such indoctrination utilized the site of Nazi victims' mass graves. Veterans and former partisans assembled for ceremonies at those sites and stated their recollections of Nazi brutalities. The political workers pointed out that imperialists had produced these atrocities and had also committed acts of brutality in oppressing national liberation movements that developed in other countries. These comparisons were used to unmask the imperialistic threat of Soviet enemies—the Americans.

Political workers in the Strategic Rocket Forces used every moment of their contact with soldiers, even appealing to the soldiers while they were on combat alert duty. Soldiers on duty in underground rocket bunkers were visited by political workers.[9]

Even in the face of new political thinking aimed at reducing the mutual military threat of the Soviet Union and the United States, the Soviets continued to instill hatred for Americans in their regular indoctrination programs. For example, a Soviet military journal included an article stating that hatred should be developed because "hate is the most important quality of the soldier and sailor since it promotes vigilance and combat readiness. Our hatred for imperialist plunder has deep roots. Our class adversaries are stained with the blood of millions of Soviet people who were killed during the wars initiated by imperialist aggressors. These imperialists are responsible for the trouble and suffering of millions of people all over the world. Our right to hate the imperialists is justified by their current blood policy and their intent to invade our homeland with weapons of mass destruction."[10]

The Soviets were developing not only hatred but also arguments to justify it and explain to the Soviet soldier why he should hate the Americans. An important reason for hatred was to overcome the Soviet soldier's interest in the American lifestyle and any sympathy he might have for such American items as jazz, cars, and freedom to travel. Hatred was used to build up a distrust of everything foreign.

During the fifties, there was little difference between the Soviet military evaluation of the American soldier and the evaluation used in propaganda. Both evaluations were summarized by F. O. Mikchei: "American soldiers are very dynamic and have good technical skills, but they lack discipline and have low moral qualities. They have only a narrow political mental outlook."[11] In the seventies, that evaluation started to change.

The majority of Soviet military writers pointed out that, in a future war, American soldiers would be well indoctrinated for fighting the Soviet Union.[12] "The future war will be very severe because the American soldier has a very strong anticommunist view. He is aggressive and ready to carry out his officers' orders. He is ready to be exposed to reasonable risks. He is obsessed with the idea that he is a superman. However, he is a selfish person, wants to make a profit, uses drugs and alcohol, and he is inclined to desert."[13]

Although Soviet military leaders changed their assessment of the American soldier, Soviet propagandists clung to their old picture of the American military gangster-murderer. Because the good fighting quality of the American soldier was recognized, propaganda was increased. Every attempt was made to show that the American soldier was actually an object of scorn. The propagandists stressed the message that the American soldier must be thoroughly hated. Such hatred produced the Nazi holocaust.

Notes

1. M. P. Korobeinikov, *Sovremenniy Boy I Problemi Psikhologii.*

2. Polkovnik B. W. Karpovich, *"Ideologicheskaya Obrabotka Soldat Armiy USA i Anglii"* (Moscow: Voyenizdat Ministerstva Oboroni, 1952), 10.

3. Ibid., 53.

4. Ibid.

5. Ibid., 61.

6. *Kommunist Vooruzennikh Sil*, no. 2, 1986, 82.

7. *Voyenniy Vestnik* 4, 1968, 119.

8. *Voyenno-Istoricheskiy Zhurnal* 3, 1987, 67.

9. *Kommunist Vooruzennikh Sil*, no. 3, 1987, 39.

10. *Communist of Armed Forces* 14, July 1986, 71.

11. F. O. Mikchei, *Atomnoye Oruzhie V. Armii* (Moscow: Izdatelstvo Inostrannoi Literaturi, 1956), 39–40.

12. Korobeinikov, 9.

13. K. Volkogonov, *Psihologicheskay Voyna* (Moscow: Voyennoye Izdatelstvo, 1984), 263.

Intelligence and Psychological Operations

13

Col Joseph S. Gordon, USAR

Intelligence is an inextricable element of PSYOP in all phases of its activities, from the planning and conduct of operations to the evaluation of its effectiveness. For this discussion, PSYOP intelligence includes that which is generated by a PSYOP organization's own research and analysis assets as well as the intelligence which is obtained from other agencies. The major areas of intelligence activities in support of PSYOP are research, target analysis, pretesting and effects analysis, and propaganda analysis.[1]

Research

Research is geared to support the mission of PSYOP, which is to create in hostile, friendly, or neutral foreign groups the emotions, attitudes, or behavior to support the achievement of national objectives. The effective PSYOP campaign requires systematic research of the highest scholarly standards. Intelligence required for PSYOP can be divided into two categories: basic and current.

Basic PSYOP intelligence consists of general information on such topics as the history, society, politics, economy, communications media, and military forces of a given country or region. Such information is kept in files of pertinent documents and a library of area studies books. When time and resources permit, the general information is digested in the production of studies that are tailored to support PSYOP. One example is the basic PSYOP study, which supplements area studies by focusing on relevant PSYOP issues, defined as highly emotional matters, deeply rooted in a particular country's history, customs, fears, or foreign policy. The issues are selected largely for their ability to support PSYOP objectives that are classified as either cohesive or divisive.

Cohesive issues exploited could help strengthen or more closely unite the total society or particular target group. Divisive issues exploited could contribute to the separation of a group from the greater society or to the disorganization of a people. The basic study should also provide information on national symbols, emotive music, or pertinent graphics that could be used in the development of PSYOP materials. PSYOP intelligence analysts also produce special studies in response to timely events or to the need for greater focus and depth on an aspect of a country or society not covered elsewhere.

Current intelligence is obtained from all possible sources and is used to update studies and files. Should a PSYOP campaign be implemented, it is essential that current intelligence be considered to ensure that the issues already identified still pertain and that the themes and materials developed for use in the campaign are still valid.

The sources of PSYOP intelligence are diverse. But most of the intelligence required to conduct PSYOP is generally obtained from unclassified sources: newspapers, magazines, books, academic journals, studies, and foreign broadcasts. Classified reports on pertinent subjects are distributed to PSYOP organizations. Such organizations may also request intelligence collection for specific items of factual information.

Liaison with other agencies outside the Department of Defense (DOD) often can produce valuable information. Some agencies that can assist PSYOP in obtaining information about local conditions in foreign areas include the Central Intelligence Agency (CIA), Department of State, Foreign Broadcast Information Service (FBIS), the Library of Congress, and the US Information Agency (USIA). USIA not only conducts public diplomacy, telling America's story to the world, but it also undertakes extensive foreign media analysis, audience research, and public opinion polls. Finally, human intelligence is important—especially to gather information about denied areas in both peace and war. The sources of human intelligence include émigrés, refugees, and prisoners of war.

No discussion of intelligence research is complete without mentioning the qualifications of the analyst. Because PSYOP demands extensive knowledge of the history, politics, society, economy, culture, and military affairs (to name a few) of a given area, research is usually a team effort. However, the individual analyst qualifications should include mastery of research and writing techniques, earned advanced academic degrees, considerable experience living in a foreign culture, and knowledge of languages pertinent to the region. The complete analyst should also be able to read the French and German languages, and should have as much information as possible on countries of the third world. Highly capable intelligence analysts are essential for a successful PSYOP program.

Target Analysis

The intelligence effort in target analysis can be divided into five elements, the first of which is to determine key audiences. Based on systematic research, one determines the major elements of a given society delineated by such factors as geography, ethnic origin, religion, race, economic status, and social position. Two basic criteria are used to select PSYOP targets. One must first consider the importance of the target to the PSYOP mission, then the probability that the target can be moved to support the objectives of the campaign. In other words, one asks whether the target has influence and whether it can be persuaded to use this influence in ways that would aid US national interests.

The second element of target analysis is to determine the attitudes of key audiences. Basic research provides intelligence on long-standing attitudes toward political, military, economic, and social subjects. This information is refined as much as possible to delineate the views of various age cohorts, social strata, and occupational groups.

The third element is an analysis of current vulnerabilities within specific audiences—in essence, an updating of the preceding element to determine the level of dissension, fear, or complaints at the time of the PSYOP campaign. When properly assessed in a timely manner, such fears and anxieties can be exploited as target vulnerabilities in a PSYOP program.

The fourth element is a decision regarding message content and means of communication. The vulnerabilities and attitudes of the target provide a foundation on which to build thematic materials for a PSYOP campaign. These materials must support the plans and policies of the campaign, and the vulnerabilities/attitudes they are based on must be an accurate assessment determined by scientific analysis and evaluation. Further, it is essential to determine which means of communication (e.g., radio, television, leaflets, face-to-face) are most capable of reaching the target.

The fifth element of target analysis, testing the contents and measuring the effects of a PSYOP message, is an extensive one. It requires a separate discussion.

Pretesting and Effects Analysis

The pretesting of PSYOP materials is a crucial phase of a campaign. It is good to determine in advance that the message is clear and potentially effective. It is also good to avoid the small errors, which are easily made, in idiomatic language or cultural incongruities that can immediately destroy credibility. There are several methods for pretesting PSYOP materials, many of which can also be used to analyze the effects of communication. But testing is usually highly problematic in that the targets of the campaign are generally located in denied areas. To partially compensate for this problem, prisoners of war, refugees, or other émigrés from the denied area can be utilized.

The general sample survey is perhaps the best systematic method for determining the effects of media content. It can be used before, during, and after a PSYOP campaign. The survey involves asking significant questions of a relatively small sample of scientifically selected persons to ensure representativeness. This technique determines whether the message is understood, credible, evokes the desired response, or has provoked undesirable effects.

Another method of testing PSYOP material is the panel, which differs from the survey in that the panel participants are repeatedly questioned at regular intervals in the course of the campaign. There are at least two variations of this technique.

One variation involves choosing a small group of people to represent some larger population segment. The success of the representative panel is, of course, dependent on the selection process. Success also requires rather free access to the audience. The second variation attempts to compensate for the lack of access by assembling a group of knowledgeable panelists to discuss the merits of the PSYOP materials in question. The discussion panel has a weakness, however: significant views can be suppressed by peer-group criticism.

A final method of testing is the individual interview. Ideally, it should be performed by a trained psychologist. The main purpose of the individual interview is to obtain insights into the deeper meaning of events and to clarify the mechanisms by which these meanings are formed, perpetuated, and changed.

Measuring the effects of a PSYOP campaign is a most complex and frustrating task. One is reminded of the Madison Avenue executive who lamented that half of his firm's advertising was effective, but he did not know which half. The above-mentioned techniques of analysis can also be used in an attempt to determine the effects of a campaign. There are also two other indicators, the first of which is physical response to the campaign. If the audience acts as the PSYOP message suggested (e.g., to defect, surrender, or cause a disturbance), there is evidence of effectiveness. One must be aware, however, that this action could have been caused entirely or in part by other factors. The other indicator of PSYOP effectiveness is through analysis of the opposing media. By monitoring the press, radio, propaganda, and other documents and publications, one can find evidence of the opposition's response to our campaign.

Propaganda Analysis

Analysis of propaganda is most valuable for a number of purposes. Examination of the opposition's domestic propaganda—that directed at its own population—can be exploited in two ways. First, it can reveal general intelligence about the political, economic, social, and military situation in a given country (if not the possible intentions of action in a given situation).

207

Second, domestic propaganda in revealing such information may also disclose vulnerabilities that can be used in our PSYOP campaigns. Further analysis of external propaganda directed at our side can serve as a basis for our own counter-propaganda activities as well as a basis for our public information efforts to defend against the effects of opposition propaganda or our own population or armed forces.[2]

Regardless of the purpose or kind of propaganda, one focuses basically on seven questions in its analysis. What is the source? What is the objective? Who is the target? What attitudes are being exploited? What techniques are used? What media are employed? How successful is the propagandist in accomplishing the objective?

Although these analytical questions may seem rather simple, answering them can require sophisticated examination of the propaganda materials. It is not sufficient to merely deal with the themes and the logic conveyed in the propaganda message; one must pay much attention to the psychological meaning of the total package, including its symbols, use of language, and other emotional accouterments. One should be thoroughly familiar with the myriad of propaganda techniques (one author listed 77 different ones). Our counterpropaganda and public-information efforts must deal with the techniques as much as the themes of opposition propaganda if they are to be effective.

Although good intelligence is important in all military operations, it seems even more important to PSYOP. It is critical throughout the spectrum of operations from the initial planning to the campaign evaluation. PSYOP is not only a user, it is also a producer of intelligence. It is capable of contributing to the overall national effort as well as servicing its own needs.

Notes

1. A valuable source of this paper was Ronald D. McLaurin, ed., *Military Propaganda: Psychological Warfare and Operations* (New York: Praeger, 1982). Especially useful was Part IV: "Intelligence and Research," which included among other pertinent articles the following: Philip P. Katz, "Intelligence for Psychological Operations," and "Exploiting PSYOP Intelligence

Sources," 121–54. Also very useful was Department of the Army Field Manual (FM) 33-1, *Psychological Operations*, 1979.

2. In addition to FM 33-1 cited above, the following works were consulted for propaganda analysis: Leonard W. Doob, *Public Opinion and Propaganda* (New York: Henry Holt, 1948); D. Lincoln Harter and John Sullivan, *Propaganda Handbook* (Media, Pa.: 20th Century Publishing, 1953); and William Hummel and Keith Huntress, *The Analysis of Propaganda* (New York: William Sloane, 1949).

Bibliography

I. Quick Studies of Foreign Countries

Foreign Area Studies Division, Special Operations Research Office. *Area Handbook.* Washington, D.C.: The American University. Best one-volume survey of individual countries. Contains extensive bibliography. Recently taken over by Federal Research Division, Library of Congress.

Skye Corporation. *The World Today Series.* Washington, D.C.: Stryker Post Publications.

United States Department of State. *Background Notes.* Washington, D.C.: Bureau of Public Affairs. Useful handbooks of regions of the world with regional and individual country chapters. Covers all parts of the world. Updated each August.

II. Intelligence

Bibliography of Intelligence Literature; A Critical and Annotated Bibliography of Open Source Literature. 8th ed. Revised. Washington, D.C.: Defense Intelligence College, 1985.

The Secret Wars: A Guide to Sources in English. Vol. A, B, C. Santa Barbara, Calif.: Clio, 1980. Extensive bibliography includes categories of covert operations, terrorism, espionage, and psychological operations.

Zell, Stanley. *An Annotated Bibliography of the Open Literature on Deception.* RAND Note N-2332-NA, December 1985. Contains many references on PSYOP.

III. Guides to Periodicals

Book Review Digest. Useful to evaluate works consulted in research.

Public Affairs Information Service. Guide to scholarly journals such as *Foreign Affairs* and *Orbis.* A similar service is provided by the *Social Science Index,* not listed here.

Readers' Guide to Periodical Literature. Guide to articles in many popular publications such as *Time* and *Newsweek.*

IV. Military Periodicals

Air University Library. *Air University Library Index to Military Periodicals.* Maxwell AFB, Alabama.

The Military Press. *Current Military Literature.* Oxford, UK. Valuable Service. Includes foreign language publications.

Quarterly Strategic Bibliography. Good on military affairs but also considers broader aspects of power.

V. Foreign Broadcast Information Service

Transdex Index. Guide to the FBIS Daily Reports and JPRS Translations (see below).

VI. Newspapers

The New York Times Index. Guide to newspapers. Considers itself the "Book of Record" for the US.

VII. Various

Chiefs of State and Cabinet Members of Foreign Governments. Updated monthly.

Current News. Pentagon clipping service of US newspapers on defense-related issues. Distributed daily in the Washington, D.C. area.

Foreign Broadcast Information Service. *Daily Report.* Newspaper and radio broadcasts translated into English each day.

_____. *JPRS Special Translations.* Translations of lengthier pieces or groups of clippings on given subjects from foreign sources.

VIII. Style Guide

College of Aerospace Doctrine, Research, and Education. *AU Press Style Guide.* Maxwell AFB, Ala.: Air University, 1994.

Turabian, Kate L. *A Manual for Writers of Term Papers, Theses, And Dissertations.* 4th ed.

What the US Information Program Cannot Do

George V. Allen

An international information program cannot change the basic attitudes and opinions in foreign countries.

One of the great American fallacies is the notion, prevalent among people in all walks of life, that we need to explain ourselves, our policies, and our way of life to foreign peoples so they will love us—or at least will understand and sympathize with our point of view. I submit, however, that this point of view is not realistic and that those in the academic world and other professionals in communications should be tough-minded enough to face certain facts squarely and realistically.

While I was director of the United States Information Agency (USIA), I was often asked by congressmen to explain why the "Voice of America" (VOA) seemed to have difficulty in getting the American story across to the people of foreign countries. It should be very simple, I was told. You need to explain that the American way of life—including our democratic principles, our respect for human rights, and our private enterprise—has produced the highest standard of living in the world. Everybody admitted that not only the upper strata but the common man in the United States had more of the good things in life—more shoes, clothes, leisure time, music, vacations, and opportunities for advancement than people of other countries. Why can't you just keep pointing that out to them on VOA? The job should be easy.

However, Uncle Sam was regarded as Mr Rich. We were presumed, with some justification, to want things to remain

Excerpts from John Boardman Whitton, ed., "What the U.S. Information Program Cannot Do," *Propaganda and the Cold War* (Washington, D.C.: Public Affairs Press, 1963). Reprinted with the permission of the *Public Affairs Press*, copyright holder.

more or less as they were. We talked a good deal about evolution instead of revolution, and we made it clear that we did not want the government of Cuba to seize American-owned sugar refineries or other American property without adequate and prompt compensation. We insisted on reforms (e.g., Latin America under the Alliance for Progress) but we wanted them to be instituted by orderly legal process while the miserably poor wanted to turn the world upside down overnight. They regarded the United States as basically in favor of the status quo. (All rich people are supposed to be that way.)

More significant, perhaps, is the fact that Moscow was regarded by most of the poor people around the world as the friend of the poor and of the rebel. When one asks how it is possible that so many Cubans were attracted towards Moscow rather than the United States, the painful but realistic answer is because they thought Moscow was more likely to support them than we were. Demagogues such as Fidel Castro or Juan D. Peron, who depended for their chief support on the rabble, were likely to shout defiance at Uncle Sam—and their followers were likely to cheer them for it.

The "Voice of America" could not change the basic fact that the US was rich while most people in the world were poor. The more we talked about our high standard of living, the fewer friends we had.

What is the answer? For those of us concerned with communications, we must recognize the facts of life. We must try to see ourselves as others see us. We must understand *their* motivations and reactions as we try to help them understand ours. We have put too much emphasis on explaining our point of view and not enough on understanding theirs.

Let us turn now from economic to political matters. One of the principal foreign policies of the United States has been the decision to participate in and foster collective security for the nations of the free world. The purpose of the USIA, in the words of President Eisenhower shortly after the agency was established in 1953, "shall be to submit evidence to peoples of other nations, by means of communication techniques, that the objectives and policies of the United States are in harmony with and will advance their legitimate aspirations for freedom, progress, and peace." In other words, we must make our policies as convincing and acceptable to foreigners as we can, presuming that our policies *are* in harmony with their proper aspirations.

I was in India when the United States gave military aid to Pakistan under our collective security policy. The archangel Gabriel could not have made this palatable to the Indians or convinced them that it was compatible with their aspirations— legitimate or otherwise. There was no policy we could have adopted that could have convinced both sides we were their friends.

The heavy responsibilities of the United States sometimes require us to take positions that please nobody. There is danger in expecting too much of communications techniques. I was once told that only the VOA could win the Berlin dispute!

There is a tendency for college professors to claim too much for the growing field of communications (sometimes improperly called psychological warfare). Many universities are rapidly developing studies, and even faculties, in this specialized field. But if those in the academic world and we in government overstate our case for communications, we are likely to make trouble for ourselves. Propaganda can do little to remove the basic problems of the have-nots, the national rivalries of Pakistan and India, or the racial animosities of Africa. And whatever communications can do will take a long time. Like education, communication is not likely to avoid takeovers in third world countries.

We must put forward an honest, objective, and truthful information program and then make it available in comprehensible terms to as many people as possible by the most effective media. Then we must rest our case with the common sense of mankind. I suppose one must have a mystic faith, as Jefferson did, in the ability of the common man to make a right decision when given adequate information and freedom of choice. If you do not have this faith, I doubt that you should be in the communications field.

Let me repeat, however, that we must be realists. Nations will be saved from aggression by a combination of forces, including political, economic, psychological, and military—the latter being possibly the most significant in our present sad state of international chaos.

Defector Operations

Lt Col John Ozaki, USA

Defector operations must be carefully planned and coordinated at all levels.

Because of the economic and political instability that prevails during internal defense operations, many insurgents can be persuaded to return to their government's cause if a sound defector plan is instituted. The successful campaign against the Huks in the Philippine Islands and the more recent Chieu Hoi (open arms) defector program in Vietnam tell us it is possible to formulate a defector program for internal defense and development. Success, of course, depends upon proper implementation.

The successful defector program requires national coordination, and it should have the objective of supporting the existing government. Specifically, that objective may be met by any of the following:

• Inducing the maximum number of insurgents to discontinue voluntary support of the insurgent program and to support the legally constituted government.

• Exploiting for intelligence and psychological operations those individuals who have returned to the side of the government.

• Fulfilling promises to defectors by providing to them and their families security and economic support, to include vocational and job opportunities that help them become self-supporting.

Reprinted by permission from *Military Review* 49 (Fort Leavenworth, Kans.: Army Command and General Staff College, March 1969), 71–78.

- Enlisting the defectors for specialized jobs and units where their knowledge of the enemy's techniques can be utilized.

To increase the chances of a successful defector program, US representatives to the host country at the national level should seek approval of the following policies, especially in cases where the United States is advising a host country on military matters:

- The host country should establish an agency to be responsible for execution of all aspects of the defector program, which should be equal in status to major components of the government such as the military departments and the national police. This agency's organization should function on the principle of centralized direction and decentralized execution from the national to the lowest level.

- US agency responsibilities and policies for defector operations must be clearly established from the national level down to receipt of a defector by a combat unit.

- Combined agreements should be made whereby a review of performance can be accomplished for the purpose of replacing ineffective administrators.

- The responsible US agency should have control of funds provided by the United States that are used in direct support of the defector program.

- Combined civilian and military counterparts should be established at each level where major defector operations take place. These would include district, sector, regional, and national levels as appropriate to the territorial organization of the country.

The objective of the inducement program is to prepare the members of the insurgent forces to quit their cause and join the legitimate government. The overall effect lets the insurgents know that the government is aware of their plight and wants to forgive and welcome them back. Closely associated with this program is the requirement of informing loyal citizens and military organizations about the program. These are the groups that play an important role in the rehabilitation of the defectors. The loyal citizens must accept the defectors back into the society or the defectors will probably return to the insurgent side.

Inducement operations are the responsibility of PSYOP personnel although intelligence organizations also play a major role in this part of the program. The intelligence community provides PSYOP personnel with information gained from interrogations designed to determine why insurgents have quit their cause. If feedback provided by intelligence is timely and valid, this information can be used to induce more insurgents to quit their cause. Intelligence organizations benefit from this type of mutual support because, with more insurgents defecting, there is the probability of increased intelligence.

Themes

Before a PSYOP program is implemented, a careful analysis is made to determine the vulnerabilities of the insurgents. Common vulnerabilities of insurgent forces are hardships, disillusionment because of slow progress, and fear of getting killed. To be effective, the term *hardship* must be translated into meaningful facts such as insufficient medical services, low pay, and long family separations. The success of any PSYOP effort depends on close coordination with intelligence agencies.

Maximum use of radio, loudspeakers, newspapers, leaflets, and other publications will be the mainstay in the dissemination link. The PSYOP operator, however, constantly seeks new and unusual techniques to spread the word of the government concerning the promises of the defector program. Innovations such as badges, postage stamps, and imprinted balloons publicize the defector program. Encouraging local government officials to speak about the program may also help.

PSYOP Utilization

Former insurgents are used to the maximum extent possible in inducement operations. They are extremely effective in developing PSYOP material because they know the environment and the modus operandi of the insurgent. If defectors are integrated into the PSYOP effort, their knowledge of the habits, customs, and idiom of the insurgents will make the PSYOP effort more effective. Additionally, these individuals are used in evaluating PSYOP material before it is disseminated.

Small units of armed defectors are organized to perform propaganda missions in locally contested areas. Such units are effective because they speak with firsthand knowledge. For security purposes, selected individuals of known loyalty are incorporated in these special units.

An effective system of rewards is one of the most important aspects of a defector program. Rewards are coordinated to ensure that each defector is paid for the assistance he provides to the government and that the pay is equitable. A central office monitors all rewards and establishes a well-publicized standard scale to ensure that rewards are in line with those previously paid. Defectors are questioned to determine whether they have received all rewards due them.

Speed in making payment is necessary so that the impact of the deed is not lost; consequently, funds should be made available to local officials so that rewards can be paid immediately. When security permits, rewards should receive wide publicity so all can see that the government lives up to its word. This added emphasis may induce other defections, particularly when large sums of money are involved.

Exploitation of the returnees involves obtaining information, disseminating the intelligence derived from this information, and using the intelligence to defeat the enemy. The following major considerations concern the exploitation of defectors:

• Treatment promised to returnees must be delivered. Initial reception is particularly important because it has an important bearing on how much an individual will contribute to the government. A returnee who is properly treated may even volunteer for intelligence and PSYOP roles. On the other hand, an individual improperly treated becomes an easy target for reindoctrination by antigovernment forces.

• Qualified interrogators are available at the lowest level possible, which is important because of the insurgent's characteristic of frequently moving. It is imperative that defectors be interviewed within the first few hours after defection to determine their knowledge. Interviewers also need the necessary communications to disseminate information to responsive units. When the number of qualified interrogators is limited, consideration should be given to pooling and dispatching them by air as the situation requires.

- Security must be provided for both the defector and the defector's family. Insurgent forces are likely to place maximum effort on retaliating against a defector (and the defector's family) to discourage other defections.

- Clearly defined policies on exploitation must be established.

One of the major problems confronting a commander is that many of his officers and men will not trust the returning defectors. To counter this logical distrust, the commander implements a massive command information program that includes handling, treatment, and use of defectors. Experience proves that one of the best ways to reverse the feeling of distrust is to publicize the help that defectors give to the friendly forces. The returnee's knowledge of the insurgents' tactics, terrain, and current situation is invaluable to the countering forces.

One of the government's most difficult tasks is to identify the enemy and the defector is often the only person who knows the enemy; consequently, a major task of friendly military personnel is to exploit this knowledge. Former insurgents may be used as scouts, guides, members of collection and reconnaissance units, and monitors for interpreters.

Special interrogation procedures are necessary for working with a defector. Experience has shown that most defectors will talk freely without the use of pressure. Generally, the best technique is to employ the interview method. The interviewer first must gain the confidence of the defector, and interviews should not be continued unless the interviewer believes that good rapport and communication have been established. Because most defectors are highly apprehensive immediately after their escape, it often requires several interviews to establish communication. It may be useful to have earlier defectors talk to all new returnees, telling them that they have made the right decision by defecting.

The interview environment is extremely important. Although a private room is desirable, elaborate equipment is not necessary. With new defectors, it is helpful to have intelligence requirements mentally fixed so that resort to paper and pencil is not necessary. The defector should be convinced that he is important and has valuable information that can be used to help defeat the enemy.

Rapid and accurate interrogation of knowledgeable returnees is necessary if effective exploitation is to be accomplished. Generally, the best procedure is to use normal military communications to report information gained from interrogations. Communications between police, government officials, and other sources are coordinated because many insurgents defect to governmental agencies other than military units. Reporting by those agencies helps to speed information to the user.

If the country is so organized, there is a military reporting headquarters at district, province, and region. Forms used for reporting must be simple, and some basic information should be reduced to blocks for check marks. It often will be beneficial if the forms are bilingual.

Intelligence agencies keep the lower echelons informed of any special requirements so that special reports can be forwarded. Each echelon in the reporting channel keeps a reference file on all defectors. The file contains basic information, such as date of rally, area of operation, former job with insurgents, special knowledge, and disposition. The information is readily available so that an individual can be quickly located for exploitation at any time.

Security

Security for defectors is essential. If the exploitation program is working well, the insurgents will counter it with attacks against individual returnees and compounds where returnees are kept. Normal active and passive security measures are improved by organizing the defectors into self-defense units. Special protection is given to high-level defectors—perhaps evacuating them from the immediate battle area even though they may be able to provide exploitable tactical information.

Special consideration must also be given to a defector's family. Immediately after a person defects, the family should be located. If they are in an enemy controlled area, their identity may have to be concealed before the defector can be exploited. When defectors cannot be exploited because of the family's location, plans should be made to evacuate the family to a secure area.

A successful defector program includes effective means for returnee rehabilitation, a phase that requires national support. Planning and coordination of national resources are required when jobs, arable land, and educational facilities are to be provided.

Reception Centers

Rehabilitation begins at the reception center as soon as a returnee has been exploited for intelligence and PSYOP. Centers are established in secure areas near the defectors' homes, if possible. It is of utmost importance that all promises disseminated by the PSYOP program be fulfilled as soon as possible.

A new returnee is made to feel welcome. Experience has shown that an effective way to help a new returnee overcome initial shock is to assign him a sponsor—a returnee who has been at the center for some time. A special ceremony (such as a retreat formation) is held to introduce this individual to the rest of the group. In addition to temporary jobs, such as sponsoring members, returnees are hired to fill as many of the permanent administrative jobs at the center as possible. They know the problems of the returnee; therefore, they are very effective in rehabilitating the former enemy.

While at the centers, individuals are given reindoctrination courses on government objectives and the responsibilities of individual citizens. Detailed programs of instruction are centrally prepared, and supporting reference materials are provided to the rehabilitation administrators.

Vocational training may be centralized because of limited facilities and qualified instructors. This training is on a voluntary basis, and national coordination is required to ensure that those individuals who successfully complete vocational training become gainfully employed.

To assist in their own protection, defectors should be organized into self-defense groups. The rehabilitation complex should be tied into the defense plan of the area, with some government forces placed in the center or located close by. Within the center, planted returnees are used to determine whether the enemy is trying to infiltrate the program. These specially trained individuals determine whether the center is providing the defectors all benefits due them. They also evaluate the center's overall effectiveness.

The defector program must include a follow-up phase to determine whether the defectors are remaining loyal to the government and becoming self-supporting. Field representatives working with the police receive information from defector program administrators when defectors have moved into their areas. These representatives determine what problems, if any, the defectors are having in being accepted by the general population. As a minimum, a system must be established whereby a former insurgent is required to register with the police in the area where he resides. The police would then be responsible for the security and surveillance of these individuals.

Returnees whose home areas are under insurgent control may require temporary homes. It is not advisable to place them in defector villages because they then become easy targets for the insurgents, which makes it harder for them to be accepted by the general public. Defectors resettle where they can find employment in trades learned during the rehabilitation portion of the program.

History demonstrates that a sound defector program is a valuable aid in defeating an insurgent movement. A good defector program supports the existing government by reducing the number of active insurgents, gaining support from the general population, and providing information that can be exploited to defeat the enemy. A good defector program is another form of combat power. To neglect it invites a longer conflict and unnecessary casualties.

Tactical PSYOP and Strategic Objectives

Ronald D. McLaurin

Exploitation of tactical opportunities for strategic PSYOP can be crucial to achievement of long-term goals.

A major cause of disaffection from the government during the early years of the Hukbalahap insurgency (1946–64) in the Philippines was the mistreatment endured by the populace at the hands of Philippine security forces. Accounts of individual abuses by police powers in the Philippines are legion. From independence until 1950, the poorly trained security forces suffered from bad morale and worse discipline. In the field, troops typically foraged for supplies at the expense of poor peasants. Indiscriminate weapons firing resulted in unnecessarily high noncombatant casualties. Worse, the Philippine constabulary sometimes shelled whole villages when some residents were suspected of harboring or abetting Huks. Suspects were seized without due process of law, then beaten or otherwise treated inhumanely. Police corruption was widespread and resulted in such bitter, frustrating experiences for the peasants that many of them supported or joined the Huks.[1]

After Ramon Magsaysay became secretary of national defense in September 1950, however, many important and effective steps were taken to protect noncombatants during tactical operations and to build confidence in their military forces.

The first step, an innovation in tactical organization, improved the supply situation in the armed forces. (Inadequate supplies had led to foraging by troops in the field.) The introduction of self-sufficient battalion combat teams (BCT) reduced the interdependence level, which had slowed army responsiveness. The BCTs displayed initiative, flexibility, and resistance to Huk intelligence-gathering. BCTs were "autonomous in almost every sense except for strategic command, which was placed under the commander of the four military areas in the country."[2]

Second, in those cases where security forces abused their authority or did not treat civilians with respect, an effort was made to ensure that prompt and appropriate punishment was meted out. Because the objective was popular sympathy, not merely appropriate punishment, a certain degree of publicity attended these self-administered acts of justice on the part of the military.[3]

Third, civil affairs officers were attached to each unit to contact and work with the people—to inform them of the government's objectives and to make the military aware of the needs, interests, and problems of the local citizenry. This activity was designed to overcome the troops' previous lack of positive contact with the Filipino masses. It resulted in close relationships, more effective civic action programs, and better tactical intelligence. Civil affairs officers were also responsible in large measure for the implementation of the population protection program.[4]

A further instrument to ensure that the armed forces acted with self-restraint toward the civilian population consisted of an arrangement with the Philippine telecommunications facilities whereby any citizen, for the equivalent of five cents, could send a telegram of complaint about troop behavior directly to Secretary Magsaysay's office. The secretary maintained a pledge to respond to each complaint and initiate an investigation into the circumstances that gave rise to it within 24 hours. The availability of this means to redress grievances, and the government's determination to follow through, were impressed upon the armed forces for maximum deterrent effect.[5]

Yet another change involved the dispatch of military doctors with individual units in the field. This function served both the military and noncombatants in the area. The doctors provided medical aid to civilians who were injured through either combat or accident. They also gave general medical attention to people who in many cases had never seen a doctor before.[6]

As the program to protect and work with the indigenous population progressed, even more flexible approaches were essayed in an effort to ensure the protection of civilians from abuse, loss, or casualty during tactical operations. By May 1954, "the patrols were firing at the Huks only as a last resort."[7] Through these and other initiatives, PSYOP at the tactical level contributed to the major strategic objective of the government—reacquisition of the loyalty and support of the population.

Notes

1. See, for example, Richard M. Leighton, Ralph Sanders, and Jose N. Tinio, *The Huk Rebellion: A Case Study in the Social Dynamics of Insurrection* (Washington, D.C.: Industrial College of the Armed Forces, 1964), 29, 33, 35, 58; Albert Ravenholt, "The Philippine Republic: A Decade of Independence," in *Britannica Book of the Year* (Chicago: Encyclopaedia Britannica, 1957), 51; *The Philippines: A Young Republic on the Move* (Princeton, N.J.: D. Van Nostrand Company, 1962), 79–80; Carlos P. Romulo, *Crusade in Asia: Philippine Victory* (New York: The John Day Company, 1955), 87, 124, 130–31; Carlos P. Romulo and Marvin M. Gray, *The Magsaysay Story* (New York: The John Day Company, 1965), 109, 123; Alvin H. Scaff, *The Philippine Answer to Communism* (Stanford, Calif.: Stanford University Press, 1955), 28, 35, 36, 49–62, 119; Frances Lucille Sterner, *Magsaysay and the Philippine Peasantry: The Agrarian Impact on Philippine Politics*, 1953–1956 (Berkeley and Los Angeles: University of California Press, 1961), 26, 224 (n. 12); Napoleon Valeriano, "Military Operations," in Counter-Guerrilla Operations in the Philippines, 1946–1953: A Seminar on the Huk Campaign held at Fort Bragg, N.C., 15 June 1961, mimeographed), 29; and Napoleon D. Valeriano and Charles T. R. Bohannan, *Counter-Guerrilla Operations: The Philippine Experience* (New York: Frederick A. Praeger, 1962), 98, 133.

2. J. Gualberto Planas, "One Year of Secretary Magsaysay," *Philippine Armed Forces Journal* 4 (August–September 1951): 14; and Valeriano.

3. Republic of the Philippines, Executive Order 113 April 1950; Alfonso A. Calderon, "Philippine Constabulary," *Philippine Armed Forces Journal* 3 (September–October 1950): 4; Romulo, 131; Romulo and Gray, 124, 127; Scaff, 36; Robert Ross Smith, "The Philippines, 1946–1954," in D. M. Condit, Bert H. Cooper, Jr., et al., *Challenge and Response in Internal Conflict*, vol. 1, *The Experience in Asia* (Washington, D.C.: The American University, Center for Research in Social Systems, February 1968), 496; and "The State of the Peace and Order Campaign," *Philippine Armed Forces Journal* 4 (January–February 1951): 12, 13.

4. Valeriano and Bohannan, 106, 211–13, 221.

5. Ibid.; Charles Wolf, Jr., *Insurgency and Counterinsurgency: New Myths and Old Realities* (Santa Monica, Calif.: RAND, July 1965), 22.

6. Charles W. Thayer, *Guerrilla* (New York: Harper & Row, 1963), 41.

7. Scaff, 135.

PART IV

Case Studies
of
PSYOP Applications

Introduction

These case studies present and clarify PSYOP goals, roles, and methods. Col Benjamin F. Findley, Jr., USAFR, condenses and analyzes US and Vietcong PSYOP in the Vietnam War. He analyzes the Chieu Hoi campaign, the vulnerabilities of the North Vietnamese army and the Vietcong soldier, and summarizes the key US and Vietcong PYSOP appeals and techniques.

MSgt Richard A. Blair, USAFR, and Col Frank L. Goldstein, USAF, present the Iraqi propaganda network used during the Persian Gulf War. They explore Iraq's skill in creating and using a variety of media outlets, internally and externally.

Col Dennis P. Walko, USA, details tactical and consolidation PSYOP activities in Operations *Just Cause* and *Promote Liberty* in Panama. He presents major PSYOP lessons learned.

Laurence J. Orzell shows that Poland's underground media were nongovernmental PSYOP in action as well as a significant part of the Polish sociopolitical scene. According to Orzell, this underground media enabled the Poles to resist the enforcement of the adoption of communist ideals and values.

Colonel Goldstein refers to the 1986 Libyan Raid as a PSYOP effort and reports the results.

Maj James V. Keifer, USAF, Retired, discusses national anti-drug policy and relates the role of military psychological operations in solving drug problems.

The late Gen Richard G. Stilwell, USA, Retired, discusses the importance of the political-psychological dimensions of conflict and insurgency. He explores, through several examples, the failure of the American public, the media, and even our bureaucracy, to recognize and counter political and psychological warfare conducted against US policies.

In Major Keifer's second piece, he gives an overview of the psychological dimension of the Persian Gulf War relative to the diplomacy process and the quick end to the war.

Colonel Goldstein and Col Daniel W. Jacobowitz, USAF, Retired, in case studies on Operations *Desert Shield/Desert Storm*, support the urgent need to increase US sensitivity to the psychological dimension of warfare. They not only solidify the

231

PSYOP requirement but also point out our weaknesses and future challenges.

US and Vietcong Psychological Operations in Vietnam

Col Benjamin F. Findley, Jr., USAFR

This paper reviews and analyzes selected US and Vietcong PSYOP strategies, tactics, and problems in the Vietnam War. Included are the Chieu Hoi campaign, the vulnerabilities of the North Vietnamese Army and the Vietcong, and key PSYOP techniques of both the US and the Vietcong.

Coordination of Decentralized United States Psychological Operations Agencies

The US PSYOP effort during the Vietnam War was decentralized among the US Military Assistance Command, Vietnam (MACV), the US Information Service (USIS), and the Agency for International Development (AID). Each agency was responsible for one or more aspects of the foreign information/PSYOP effort conducted in South Vietnam. The agencies "operated independently of each other," according to Ronald D. McLaurin of the American Institute for Research in Washington, D.C. In 1965 President Lyndon B. Johnson established the Joint US Public Affairs Office (JUSPAO) to integrate the PSYOP activities, to avoid duplication of effort, and to increase the overall effectiveness of the effort.

According to then 4th PSYOP Group Commander Taro Katagiri, the variety of agencies involved made it difficult to coordinate the PSYOP effort as well as to establish centralized control and direction in the Republic of Vietnam (RVN) in 1968–69. There were repeated examples of lack of coordination. Mr Katagiri recalled that on 16 February 1969 there was an airborne leaflet mission in which a total of 84,000 leaflets were disseminated without prior coordination with the area task force PSYOP officer. And US PSYOP teams sometimes appeared at hamlets on audiovisual or loudspeaker missions only to discover a province team conducting a similar mission.

Roles and Goals in Vietnam

PSYOP responsibilities grew during the war without the guidance and control necessary to accomplish the corresponding goals. The breadth of American roles and goals led to rapid growth in the size of US PSYOP elements. PSYOP goals:

1. undermine popular support for the insurgents,
2. enhance the image of the government of the RVN,
3. increase Vietnamese understanding of, and elicit support for, US policies in Vietnam, and
4. engender international support for US policy in Vietnam.

Relatively few PSYOP-experienced personnel were available, however, and those few were spread among the many groups involved in PSYOP. According to MACV, the US PSYOP mission in Vietnam included the following entities:

- the American Embassy in RVN,
- the Mission PSYOP Committee,
- the JUSPAO,
- MACV,
- Civil Operations and Revolutionary Development Support (CORDS),
- US Army Republic of Vietnam (USARV),
- the 4th and 7th PSYOP Groups,
- Naval Forces, Vietnam (NAVFORV), and
- the Seventh US Air Force.

In PSYOP Policy number 51, 28 December 1967, the JUSPAO Planning Office established the following PSYOP priorities:

1. the Government of Vietnam (GVN) image,
2. Chieu Hoi (come home),
3. revolutionary development (agriculture and improved living standards self-help projects),
4. refugee program,

5. public safety,
6. the US image,
7. GVN mass media advisory effort, and
8. telling the Vietnam story.

The Chieu Hoi Campaign

Two special PSYOP targets were the Vietcong (VC) and the North Vietnamese Army (NVA) soldiers in South Vietnam. Two Chieu Hoi operations carried out in the Delta during 1970 and 1971 proved that PSYOP and combat pressure working together could get results. The operations were Operation *Roundup* in Kien Hoa Province and Project *Falling Leaves* in Kien Giang Province. Operation *Roundup* produced hundreds of enemy defectors, according to Colburn Lovett, a USIS foreign service information officer. One PSYOP technique was to take pictures of ralliers/defectors and have them sign a simple message on a leaflet, encouraging their comrades to join the cause. Another technique was to use loudspeaker teams of former VC soldiers who were sent back into the areas of their units to speak to their comrades in the bush. Project *Falling Leaves* combined Vietnamese and US personnel working in joint PSYOP activities. Armed propaganda teams (100 percent ex-VC) made deep penetrations and extensive face-to-face communications. All possible media were used, including boat-carried loudspeaker teams, leaflet drops, radio tapes, and television appeals by former VC.

There are significant PSYOP lessons here for all commanders. During the two-month period in 1971 when intense PSYOP supplemented military action, there were 1,150 defectors, while in the six weeks before and the four weeks after the intensive PSYOP campaign, there were only a total of 211 defectors. PSYOP, if applied on an intense scale, is of significant value to commanders in securing the surrender of enemy forces. The author believes that military operations conducted without PSYOP support will not be as successful as those conducted with it.

A PSYOP project initiated by Special Forces in conjunction with a detachment of the 245th PSYOP Company, JUSPAO, and the Vietnamese Information Service (VIS) had as its objective bringing Vietnamese government presence back to the area around Duc Co in Pleiku Province. The area had slipped into the contested category. Using the Duc Co Special Forces Camp as its base, the PSYOP effort was aimed at all of the villages and hamlets within a 10-kilometer radius. This Chieu Hoi campaign, adopted in 1963, offered communist soldiers forgiveness and exoneration for temporarily slipping into the alien communist path. They could return home to their families and the just cause of the republic. The term *surrender* was not used. Medics held sick calls over a four-day period to attract the sympathetic attention of the villagers. Over 800 villagers were treated during this four-day period. Face-to-face contact allowed the representatives of the VIS to stress the theme that the VC were preventing peace while the government of Vietnam was working for peace. Valuable information concerning the popular resentment toward VC methods came to light and the team members were careful not to make the same mistakes, particularly with regard to pressures exerted to bring villagers to propaganda sessions. At the end of four days, the operation was judged to be a success, so much so that the wives of eight VC persuaded their husbands to seek amnesty as Chieu Hoi returnees.

Vulnerabilities of the North Vietnamese Army Communist Soldier

The NVA soldier in South Vietnam presented a particularly difficult target for GVN/US PSYOP aimed at inducing surrender or defection. He had a relatively high state of indoctrination, which was reinforced by a range of psychological controls that included self-criticism sessions, the three-man cell, and endless repetition of communist themes. A contributory reason for the resistance of NVA soldiers to Chieu Hoi inducements was that defection for most did not hold the promise of an early family reunion. (Defecting Vietcong were South Vietnamese who were going home.) Moreover, unlike the VC guerrilla who may be a teenager conscripted from his hamlet and sent into battle without much party schooling or political indoctrination, the NVA soldier was the product of a closed totalitarian society. He had been subjected to communist indoctrination from his earliest school days. This made him more resistant to PSYOP entreaties. Further, the NVA soldier found himself fighting in a region unfamiliar and semi-antagonistic to him, usually in relatively uninhabited areas and with little chance for contact with the civilian population.

US/GVN PSYOP messages pointed out three options: (1) to rally, take advantage of the Chieu Hoi program and quickly become a free citizen of the RVN; (2) surrender as a prisoner of war and await repatriation at the end of the war in the safety and relative comfort of a prisoner of war camp; and (3) counsel NVA soldiers to devote all their efforts to individual survival rather than getting killed or maimed for an unjust cause. Even a partial success in this PSYOP effort would contribute to shortening the war by reducing the combat effectiveness of NVA units.

The vulnerabilities of NVA soldiers did not change much throughout the war. Separation from families, the hardships of infiltrations, fear of allied arms, and, perhaps most significantly, the contrast between what they had been told by the cadre and what they actually experienced, were the major exploitable weaknesses.

The surrender program for the Northern troops (Chieu Hoi) received the greatest amount of American emphasis and money. Reception centers for defectors (hoi chanh) were built at various locations throughout the country. They usually remained in the camps from 45 to 60 days before being released and resettled. They were given rewards (e.g., money to buy food) for turning in their weapons. NVA soldiers realized that defecting might prevent them from ever going home to the North.

Key Psychological Appeals

According to Vietnam veteran and former Air Force intelligence officer Lt Col Robert Chandler, five major US PSYOP appeals were applied in Vietnam:

1. fear of death,
2. jungle hardships,
3. loss of faith in victory,
4. concern for family, and
5. disillusionment.

The aim of the fear appeal was to convince the enemy that he faced an overwhelming danger of being killed if he remained with the communists. Message themes included "death lurks everywhere," "born in the north to die in the south," and "no shelters are safe from the bombs of the B-52." Not all fear appeals were successful, however: In one appeal, fear leaflets imprinted with the ace of spades as a sign of death were dropped and decks of cards with prominent aces of spades were left along VC trails; unfortunately, Vietnamese card decks did not include the ace of spades. Another example of fear-appeal failure was the use of gruesome leaflets depicting corpses. This had a boomerang effect because the hoi chanh felt that the government was gloating over the deaths of fellow Vietnamese.

The hardship appeal reminded enemy soldiers of their loneliness, homesickness, poor living conditions, and insufficiencies in food and medical supplies. The "loss of faith in communist victory" appeal sought to convince the enemy that the Republic was winning the struggle. It emphasized battlefield losses and the number who had already rallied. The "concern for family" appeal, one of the most effective, was very emotionally based. Loneliness, nostalgia, and the desire to return home to loved ones were primary factors motivating communist soldiers to defect or surrender. The disillusionment appeal was based on the idea that the North Vietnamese soldier might be able to withstand fears and hardships as long as he was convinced that Hanoi's aims were just, but would be more willing to defect when he became skeptical of them.

Four special PSYOP techniques were employed in Vietnam: distribution of safe conduct passes, money for weapons, focus on returning home to celebrate during the Tet New Year, and armed propaganda teams composed of hoi chanh. Many PSYOP professionals believe these teams were effective because of their personal touch to the Chieu Hoi invitations.

Important Nonmilitary Considerations

There were several misconceptions about PSYOP. Taro Katagiri, former 4th PSYOP Group commander in Vietnam, believed that many people considered PSYOP a separate and distinct activity, unrelated to other functions—especially in nonmilitary situations.

PSYOP was not amply exploited to support economic programs such as rural construction, political programs, medical assistance, or humanitarian efforts. Another misconception was that PSYOP involves just verbal communications and, thus, many Americans were insensitive to nonverbal gestures, posture, signs, and physical appearance.

An individual's response to persuasive communication in any culture is based on and reinforced by his regular interpersonal relationships. An individual usually consults a member of his/her primary group, a friend or relative, before taking action in response to persuasive messages. A great number of the VC rallied/defected to the Chieu Hoi program through an intermediary who was a relative or trusted friend. Many did not defect until they received assurance from a relative or friend that the government would keep its promises concerning good treatment and other aspects of the amnesty program.

Cultural Considerations in Vietnam

For the Westerner, one of the most important elements to understand about Vietnamese culture is the completely personal basis of the society. "Truth" for the Vietnamese is not the factual statement or actual occurrence of an event; it is in the pleasantness of personal relationships. A Vietnamese's entire character lies in not giving or receiving personal embarrassment or shame. For example, it would not be culturally correct for a Vietnamese to shout at or reproach a waiter over bringing the wrong meal order.

He would either quietly request a change or simply eat the wrong meal and pleasantly smile. Understanding and progress are less important than pleasant associations. Distrust of strangers is a part of Vietnamese culture, and Vietnamese people politely keep their distances. Vietnamese answers to questions will reflect what the questioner may want to hear. A Vietnamese will avoid lengthy associations with strangers but will be pleasantly polite. The Vietnamese culture places a very high value on the family and on elder citizens.

Vietcong Techniques

According to Philip Katz, the reverse of the trusting value was used by the VC in their attempt to deter defection. His example is of a VC or NVA soldier who claims to his comrades that he was captured by the GVN or US forces and had subsequently escaped. He testifies that while a prisoner he was treated badly and that he had firsthand knowledge that the government of Vietnam did not abide by their good treatment promise.

Katz says that research findings substantiate the fact that intimate Vietnamese associates tend to hold opinions and attitudes in common and are reluctant to depart unilaterally from their group consensus. They talk frequently among themselves and establish a trusting bond of friendship upon which to primarily base their decisions. Interpersonal communication is highly valued; mass media are of secondary importance.[1]

Vietcong National Liberation Front

Douglas Pike presents the three VC PSYOP programs of dan van, dich van, and binh van. Dan van was the VC effort to develop support in areas that it controlled while dich van was the effort to develop support in GVN-controlled areas. Binh van was the recruiting program among Army of the Republic of Vietnam (ARVN) troops and GVN civilian servants. Destruction of South Vietnam's armed forces was the overriding priority for the VC; violence, armed attacks, assassinations, kidnappings, terrorist acts, and binh van were employed.

The top objective of binh van was to induce unit desertions, preferably accompanied by an act of sabotage. The next highest objective was to induce individual military desertion or civilian defection, preferably accompanied by an act of destruction or a theft of key documents. Next was to induce major and significant opposition within the military or civil service, either covertly or overtly.

Binh van PSYOP techniques included the following:

1. enunciation and constant restatement by all possible means of a liberal VC policy toward recanting military and civil servants, including prisoners;

2. wide and intensive use of terror and psychological intimidation against key officials and military units, including killing every person;

3. use of penetration agents to infiltrate and develop support within the military and civil service;

4. use of family ties and friendships to induce or coerce military personnel and civil servants to desert, defect, or covertly serve the VC;

5. tangible and intangible rewards for those who deserted or defected, (for example, the VC announced it gave $2,000 to a group of deserters in Long An Province);

6. encouragement among potential draftees to oppose the military draft; and

7. distribution of two songbooks containing 20 songs. The emotional songs were about the homeland and total victory. They contained communist ideas, praised party leadership, and presented the great qualities of the communist guerrillas. Cultural dramas were also used to promote the cause.

US PSYOP objectives in the Vietnam War were ill-fated from the beginning, primarily because of the "foreign invader" image of the United States. US psychological operations were used simultaneously as a means of achieving US foreign policy goals and as a substitute communications tool for the Republic of Vietnam to create nationalism. We misjudged the will of the communists to prolong the revolution and bring about reunification. We also misjudged Vietnamese reluctance to support any central government. In addition, we could not win the support of the US Congress to fight the type of war that needed to be fought. Instead, we fought a protracted war that was doomed. We lost the hearts and minds of the Vietnamese and our fellow Americans; we lost them to the PSYOP of the Vietnamese communists. *We must learn the values of strategic, operational, and tactical PSYOP from this devastating experience.*

Notes

1. Philip Katz, *PSYOP and Communication Theory.*

The Iraqi Propaganda Network

MSgt Richard A. Blair, USAFR
Col Frank L. Goldstein, USAF

During the Persian Gulf War, Iraq had considerable success in mounting a psychological warfare (psywar) campaign that reached a wide domestic and foreign audience. While some might attribute this success (at least in part) to overzealous international media providing newsworthy coverage of events in the Gulf, the bulk of the accomplishment is a credit to Iraq's skill in creating and using a variety of media outlets as propaganda dissemination mechanisms.

Following his established pattern, which was designed to prevent the accumulation of power within any one governmental entity, Saddam Hussein distributed propaganda duties among many offices within the government and the Baath Party. While most of these offices officially answered to the Ministry of Culture and Information (MCI), they were actually managed through informal channels by the Baath Party, the Revolutionary Command Council (RCC), and Saddam Hussein himself.

Within Iraq, all broadcast facilities are owned and operated by the Broadcasting and Television Organization of the MCI. Prior to the invasion, Iraqi broadcast capabilities included two prime-time television broadcasts, two domestic radio services—"Baghdad Domestic Service" and "Voice of the Masses" (VOM)—and shortwave radio broadcasts of VOM in Kurdish, Turkoman, and Assyrian.

Shortly after the invasion, Iraq took over Kuwait's radio facilities and began operating the Provisional Free Kuwait Government Radio program. They also implemented at least five additional shortwave radio programs designed to undermine the Saudi and Egyptian governments and the morale of Arab troops in the Gulf.[1]

Once Allied coalition bombing commenced, the Iraqi broadcasting capability was severely hampered. Iraq's television capability was completely destroyed and the only radio broadcasts noted were "Baghdad Domestic Service" and "Voice of the Masses"—both operating at greatly reduced signal strength.

The printed media within Iraq was also largely controlled by the government. The Baath Party published the largest daily newspaper in the country, *Al-Thawrah.* The MCI managed Al-Jahameer Press House, which published two other widely read dailies in Iraq, the Arabic-language *Al-Jumhuriyah* and the English-language *Baghdad Observer.* The Ministry of Defense also published an Arabic daily newspaper, *Al-Oadisiyah,* and other ministries and agencies of the government sponsored the majority of some 12 weekly papers and monthly magazines published in Iraq. Even where they did not control a publication directly, the government still influenced its contents—the MCI's General Establishment for Press and Printing licensed and censored every publication that appeared in the country.

The Iraqi government also expended considerable resources to reach foreign audiences. Iraqi television aired video clips to foreign broadcast organizations over both ARABSAT and INTELSAT. The Iraqi News Agency (INA) gathered and translated news and commentary from Iraqi press, radio, and television sources and disseminated reports by wire to subscribers around the world. The INA distributed its material even further through its participation in the Non-Aligned News Agency Pool, a service of the Non-Aligned Movement that facilitates the sharing of news items among member nations.

Iraq also controlled several other significant media resources, the most notable of which was the "Voice of the Palestine Liberation Organization," a radio program produced by Palestinian elements in Baghdad and broadcast over the Iraqi shortwave radio network. They had varying degrees of influence over some ideologically compatible media like the Libyan news agency JANA and the Palestine Liberation Organization (PLO) radio in Algiers. By mid-November, Iraq was successful in rallying information officials of Palestine, Tunisia, Jordan, Yemen, Mauritania, and Sudan to issue a joint statement endorsing the Iraqi position and discussing ways of bolstering media cooperation to confront the intensive hostile media campaign launched by the United States and its allies.

Additionally, Iraq took advantage of every opportunity to provide its views to foreign news agencies through government officials, both at home and abroad. For example, an informal examination of three days' worth of Iraqi propaganda in mid-August revealed official statements made to the press in various countries by three RCC members, seven government ministers, 10 ambassadors, a governor, a military commander, and the speaker of the National Assembly.

Iraq also displayed great success in varying the spokesmen mouthing their propaganda. Sometimes this simply took the form of issuing statements in the name of various groups or individuals; for example, the children and vanguards of Iraq or the American hostage Golden Johnson. More often, it consisted of quoting pro-Iraqi statements elicited from various amenable personalities such as Palestinian leaders George Habash and Nayif Hawatimah or of distorting statements by world leaders whose condemnations of Iraq left room for misinterpretation.

Iraq allowed the first Western journalists into Baghdad on 18 August 1990. Later, however, all foreign journalists except a news team from Cable News Network (CNN) were expelled— and these journalists were always strictly controlled. They were escorted everywhere, and all news reports leaving Iraq were censored by the government. Although most of the world knew that Iraq was censoring these journalists' reports, use of this link to the CNN network still greatly enhanced many of Iraq's propaganda campaign objectives.

Additionally, Iraq used covert methods to access outlets that could not be influenced openly or directly. Stories of wide-spread Iraqi bribery surfaced continuously throughout the crisis. Supposedly, influential members of various countries were offered valuable gifts and monetary compensation to promote pro-Iraqi sentiment. One confirmed report of this activity resulted in the expulsion of an Iraqi Embassy official in Pakistan for "providing financial assistance for the publication of propaganda materials against the state [of Pakistan]."[2]

Iraq also financed several newspapers in Western Europe to gain Arab support throughout that region. However, toward the end of the crisis, several of these newspapers either closed down or reduced operations. It is uncertain whether Iraq gave up this project due to budgetary reasons, lack of discernible positive results, or a combination of both.

Iraq's counterpropaganda efforts completed the network. With some modifications, Iraq took the approach that nearly all hostile themes and attacks should be countered openly and vociferously. Therefore, the Iraqis collected hostile propaganda, analyzed it, developed appropriate themes in response, and inserted them into their own campaigns within hours. For example, within hours of President George H. Bush's address to the US Congress on 11 August 1990, INA carried a lengthy rebuttal and sent it out over the newswire. Shortly thereafter, INA carried a second long rejoinder composed by Foreign Minister Tariq Aziz. By the evening of the 12th, the editors of every Iraqi daily paper had prepared, for publication in the morning papers of the 13th, their own attacks on the speech. To complete the cycle, INA then compiled a summary of the newspapers' criticisms that was circulated on the newswires early on the 13th.

In analyzing Iraq's success in disseminating propaganda to foreign and domestic audiences, three key components are easily identifiable. First, Iraq's invasion of Kuwait grabbed world interest and focused international media. Second, the government's skillful use of modern technology in mass communications provided it with instant access to the world populace. And third, Iraq used aggressive and imaginative propaganda campaigns to obtain and retain the widest possible mass audience.

By combining these three elements, Iraq not only made significant contributions to its national objectives but also created a powerful force multiplier. This multiplier presented the illusion that the Iraqi military capabilities and will to fight were much stronger than reality proved them to be.

Iraq's achievements in coordinating their internal and external propaganda efforts resulted in significant success in reaching world audiences. Although ultimately unsuccessful, the Iraqi effort clearly demonstrates that imaginative use and integration of today's mass communications technology allows any country to influence audiences well beyond their borders.

Notes

1. These five shortwave radio programs were "Holy Mecca Radio," first observed on 10 August 1990; "Voice of the Egypt of Arabism," first observed on 11 August 1990; "Voice of the Peninsula and the Arabian Gulf," implemented on 29 August 1990; "Voice of Peace," implemented on 11 September 1990; and "Voice of Arab Awakening," first observed on 13 October 1990.

2. Pakistani newspaper *Markaz*, 16 January 1991.

Psychological Operations in Panama During Operations *Just Cause* and *Promote Liberty*

Col Dennis P. Walko, USA

US military psychological operations units figured prominently in Operation *Just Cause* in Panama. Even US television viewers gained some appreciation of PSYOP; they saw broadcasts from the former Panama Defense Force-dominated national television channel that included "mysterious" US and Panamanian nation symbols. They saw safe conduct passes dropped on Panama Defense Force and "Dignity Battalion" (DIGBAT) positions, and they heard tactical loudspeakers appealing to enemy forces to cease resistance and surrender. Perhaps our country's last television view of PSYOP activity in Panama was that of the loudspeaker in front of the Papal Nunciature blaring heavy metal rock music to "drive Noriega out of his mind" or "keep him awake."

Military PSYOP disappeared from US public view immediately after Noriega surrendered to US authorities, only to reemerge during Operations *Desert Shield* and *Desert Storm*. Americans once again were exposed to news articles featuring PSYOP leaflets and loudspeakers.

In each case, the attitude of our mass media and the ensuing public response were favorable to PSYOP. Articles in newspapers and journals portrayed PSYOP as both cost effective and critical to success. Official and unofficial US military circles said PSYOP was a valuable force multiplier that had induced enemy forces to abandon their positions or surrender, thereby reducing loss of life on both sides.

Less conspicuous was the bulk of psychological operations during *Just Cause*. Most PSYOP activities and accomplishments in Panama were hardly noticed by either the US public or the general military community. The special operations community, however, did notice. The lessons learned in Panama were incorporated into standing operating procedures. Where possible, immediate changes were made to capitalize on the PSYOP successes of *Just Cause* and *Promote Liberty*. This led to improved production, performance, and effect in the next contingency, which took place within six months after the return of the last PSYOP elements from Panama. Operations *Desert Shield* and *Desert Storm* employed PSYOP at an order of magnitude and effectiveness that many credit to the lessons learned from Panama.

Recognizing that much of *Just Cause* and its complementary civil-military restoration operation, *Promote Liberty* is still classified, this article has been prepared to informally discuss PSYOP during the Panama contingency. It will briefly cover the evolution of PSYOP plans and operations, describe their objectives, and show how PSYOP were successfully executed. Hopefully, this piece will show there was a lot more to PSYOP during this contingency than rock music and surrender appeals, and that PSYOP did, in fact, contribute significantly to the overall success of the operation.

Panama contingency plans had been under development for years prior to *Just Cause.* These contingencies were to range from evacuation of US and third country nationals to full-scale combat operations against Noriega, his Panama Defense Forces (PDF), and (later) the DIGBAT goon squads. The latter targets were increasingly viewed as threats, not only to Panamanians, but also to US dependents. Contingency planning culminated with consolidation and reconstruction operations to follow any combat scenario.

Psychological operations were well integrated into the planning process. The worst-case scenario involved combat operations in Panama. For this option, planners recognized the importance of promoting prompt cessation of hostilities and minimizing civilian casualties. The following were key concerns for this option:

- US casualties
- Panamanian civilian casualties
- Collateral damage

- Delay in US troop withdrawal
- Disinformation and hostile propaganda

The five concerns were interrelated. If the PDF offered concerted resistance to US troops, casualties on both sides would escalate, civilians would likely suffer, Panama City and US military installations might sustain severe property damage, and US troops could become bogged down in an effort which would increasingly be susceptible to counterproductive local, US, and worldwide criticism. Related to all this was the concern that the Noriega-controlled Panamanian media—essentially a propaganda machine—would continue to broadcast material detrimental to US interests and stir up doubts as to the effectiveness of our operations.

Additional concerns for consolidation or reconstruction operations included repairing the government and economic infrastructure and establishing a functioning democracy. Specific PSYOP missions and requirements were identified to support contingency plans and address all the aforementioned concerns.

On the tactical level, loudspeaker teams were scheduled to accompany all major ground combat units. Their objective was to convince the enemy to cease resistance and surrender while advising innocent civilians how to stay out of harm's way. As the plan matured over time, loudspeaker teams received increased emphasis: a few months prior to the operation, the designated commander for Joint Task Force-South (CJTF-South), then Lieutenant General Stiner, directed that loudspeaker teams be provided to each infantry battalion (Army and Marine Corps) and each SEAL battalion-equivalent participating in the assault phase of the operation. This was probably the highest loudspeaker-to-combat force ratio in the history of the US military.

On the national-strategic level, a PSYOP task force would be formed, based upon the regionally oriented 1st PSYOP Battalion (Bn) augmented with additional loudspeaker, radio, and television assets from the 4th PSYOP Group (Gp). The PSYOP task force (TF) main body was to deploy early, assume control of ongoing tactical PSYOP, and provide general PSYOP support to the operation, employing full media production and dissemination capabilities. Its objectives were to conduct national-level PSYOP designed to minimize interference and resistance, and to foster support for US military operations and efforts by the Panamanian government to restore law and order.

Various prepackaged PSYOP materials—prerecorded TV, radio, and loudspeaker tapes; radio and loudspeaker scripts; music; and designs for printed leaflets and posters—were developed from 1987 to 1989. These were approved by an interagency committee headed by the commander of US Army South.

The commander of the 1st PSYOP Battalion, 4th PSYOP Group, based at Fort Bragg, North Carolina, was the designated commander of the PSYOP TF in the eventuality of any Panama contingency. He was responsible for developing detailed contingency PSYOP plans, annexes, and products, and for coordinating all PSYOP-relevant details with USSOUTHCOM and XVIII Airborne Corps. Prior to the actual operation, often during routing trips to Panama, the commander and other members of his command visited many of the principal target locations and identified potential production facilities—a precaution against a late arrival of PSYOP production equipment.

Facilitating this planning process was the fact that successive 1st PSYOP Bn commanders served as CINCSOUTH's senior theater PSYOP officer, knew most key players on the USSOUTHCOM and USARSO staffs, and had a small liaison cell located in the USSOUTHCOM J-3 directorate. Furthermore, once the XVIII Airborne Corps commander (General Stiner) was designated CJTF for the contingency, coordination between the battalion and that headquarters (including one-on-one meetings with General Stiner) became a frequent affair. General Stiner would later assume overall responsibility for planning and commanding the operation. He took a personal interest in PSYOP, further enhancing their planning, coordination, and execution.

By March and April 1988, Noriega had increased provocative actions against US forces in Panama, having sent harassment patrols into the US-controlled Arraijan tank farm (fuel storage and distribution depot adjacent to Howard Air Force Base) and the nearby US Army ammunition storage facility. At this point, the National Command Authority ordered additional security forces deployed to Panama, with additional deployments in May 1989. Included among these forces were an additional infantry brigade, military police, and three PSYOP loudspeaker teams. These forces remained in Panama until hostilities were initiated.

The three loudspeaker teams deployed as a tactical detachment with sufficient personnel, vehicles, radios, and speaker systems to permit reconfiguration as five teams if the need were to arise. As time dragged on, soldiers and equipment were rotated to and from Panama to such a degree that virtually every battalion in the 4th PSYOP Gp had an opportunity to deploy soldiers to Panama. This would prove extremely important: by the time of the contingency, a large number of PSYOP soldiers had firsthand knowledge of the area, the equipment, the supported commanders already in Panama, and the Spanish language.

As tension in Panama grew, loudspeaker personnel became integral members of US tactical exercise and security forces. These units often became embroiled in PDF and DIGBAT-orchestrated protests during several of the increasingly frequent exercises conducted by US forces to reassert US treaty rights. The loudspeaker teams not only effectively employed prerecorded tapes which would also be useful later, during the contingency; they prepared other announcements on the spot. These encounters served as valuable training and experience for PSYOP soldiers who would support *Just Cause* and *Promote Liberty.*

Two events in Panama were to have a significant effect upon PSYOP planning for the contingency. The first was Noriega's nullification of the May 1989 Panama presidential elections. Besides the obvious worsening of Panama's internal conditions and international relations, this act gave US PSYOP a number of obvious themes for use against Noriega and his "inner circle." It also heightened the probability of a combat contingency and energized even greater attention to the contingency planning process (including PSYOP planning and preparation). From this point forward, both the 1st PSYOP Bn commander and the 4th PSYOP Gp commander spent increasing time in Panama, providing PSYOP planning expertise to

CINCSOUTH, his staff, and the US Country Team. (In fact, both officers were to return from lengthy TDY in Panama within 72 hours of the events that precipitated Operation *Just Cause.*)

The second event was the failed 3 October 1989 coup attempt by members of the PDF against Noriega. After the coup attempt, Noriega rapidly purged the PDF of its more moderate leaders and boosted training for the paramilitary DIGBATs. Prior to this coup attempt, the US combat contingency plan had focused on removing Noriega and his cronies from power while leaving the PDF institution alone; under these new conditions, it became increasingly clear that, in the eventuality of a contingency operation, the JTF would not be able to operate within those limited US objectives. Therefore, PSYOP materials that had been prepared prior to the coup required revision. The most time-consuming revisions were those associated with audiovisual products that had been prepared for possible broadcast over television.

As time passed and the situation in Panama worsened, it became clear to the 1st PSYOP Bn commander that he could no longer keep television and radio broadcast tapes up to date. To have done so would have worn out personnel of both the battalion and the 4th PSYOP Gp's Strategic Dissemination Company. Since these products were classified TOP SECRET-SPECAT, they could only be produced at night and on holidays when noncleared soldiers were off duty and away from the media production facility. Furthermore, this activity was beginning to tie up personnel who were key to other ongoing PSYOP activities in the USSOUTHCOM area of operations. Thereafter, for initial operations the PSYOP TF would depend upon the battalion's prerecorded music tracks, silent footage and logs, and written scripts that could be modified at the last minute and read "live" by PSYOP broadcast announcers. In fact, such modifications for radio and television broadcasts were made by the commander, from Panama, over long-distance telephone within hours prior to commencement of hostilities. All were used during the first 24 hours of *Just Cause*, constituting the first official US broadcasts in Spanish heard by the Panamanian population.

After General Thurman's assumption of command on 30 September 1989, the failed coup attempt that shortly followed, and General Stiner's designation as CJTF-South, conditions in Panama continued to deteriorate. US concerns for the safety and protection of US lives and property were heightened; preparation and coordination for a Panama contingency involving full-scale combat operations were intensified.

As a result, 4th PSYOP Gp loudspeaker personnel at Fort Bragg were placed on recall status for possible deployment on short notice. Loudspeaker system shortfalls were identified and additional systems were obtained from Army Reserve PSYOP units. Loudspeaker broadcast scripts in English and

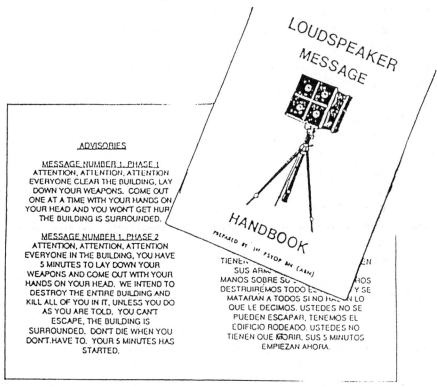

Source: 1st PSYOP Battalion, US Army

Figure 1. These bilingual booklets were carried by tactical PSYOP loudspeaker teams, along with prerecorded tapes of the same messages in Spanish. Note the messages in the extract; US troops did not mince words.

255

Spanish were refined by the 1st PSYOP Bn and published in a small handbook. Prerecorded tapes for loudspeaker broadcast were revised and additional announcements were recorded. Tapes and booklets were taken to Panama for use by the deployed teams if the time came to execute the contingency.

The 1st PSYOP Bn commander was designated by General Stiner as a member of a 20-man JTF-South staff element identified to predeploy to Panama in advance of the assault forces if the contingency were to be executed. As it turned out, this predeployment would prove important to the success of PSYOP in the Panama contingency. With the exception of the three loudspeaker teams already in Panama and the loudspeaker teams accompanying the airborne and special operations assault forces, the PSYOP TF would not arrive in Panama until after initial assault operations.

The commander, his forward liaison cell, the loudspeaker detachment, a prepositioned 4th Gp AM radio broadcast team, Volant Solo PSYOP TV/radio broadcast aircraft from the 193d Special Operations Group (SOG) of the Pennsylvania ANG, and the USSOUTHCOM J-3 PSYOP office would comprise the PSYOP planning, coordination, development and dissemination capability in Panama until D + 3. Late arrival of the main body notwithstanding, PSYOP employed a full range of media and other activities in support of initial combat operations—and with good effect.

During the early morning of 18 December 1989, two days after the fatal shooting of a US serviceman and the harassment of another and his wife by members of the PDF, the 20-man advance JTF-South staff element administratively and clandestinely deployed to Panama. The 1st PSYOP Bn commander assumed control of his forward elements in accordance with prior arrangements with USSOUTHCOM headquarters, although the members themselves still had not been informed of the pending operation, and in fact were not to be told until the night of the assault.

At 1900 hours, 19 December, the 1st PSYOP Bn commander assembled the members of his liaison element, the loudspeaker detachment commander (now the company commander who, coincidentally, was in Panama conducting his assumption of command property inventory), and the commander of the prepositioned AM radio broadcast team to notify them that *Just Cause* was to be executed on 20 December. The commander then organized the element to function on a 24-hour basis as his ministaff, interim PSYOP development cell, and radio broadcast team. Consistent with the JTF OPLAN, the loudspeaker detachment would break down into five teams which would link up with their designated supported units. The other loudspeaker teams and PSYOP liaison officers that were called for in the plan would deploy with their supported combat units and would operate in an attached or OPCON status—a command and control relationship which caused problems, as will be discussed later.

This small PSYOP force was prepared to work in shifts until the PSYOP TF arrived. Although envisioned to function only for about the first 24 hours of the operation, it was this small group which planned, coordinated, and ran psychological operations for the first three days of the contingency. Once the remainder of the PSYOP TF had commenced operations, the liaison element continued as the JTF PSYOP staff section and PSYOP TF liaison at the JTF-South headquarters, maintaining continuous coordination with all elements of the JTF and coordinating aviation requests and leaflet drops.

In addition to AM radio broadcasts, initial PSYOP activities involved TV broadcasts of prepackaged materials from the 193d SOG's Volant Solo aircraft. The recently revised scripts were read live by PSYOP soldiers who had been picked up at Pope AFB, North Carolina, while the aircraft was en route to Panama. These broadcasts—all in Spanish—notified the Panamanian population of US intent and advised how to avoid accidentally becoming a casualty. (These were the "mysterious" broadcasts that puzzled some of the US media in Panama.)

The TV channel used for broadcasts by Volant Solo was Channel 2, the national channel commandeered by the Panamanian military shortly after the 1968 coup d'état and controlled by the military until 20 December 1989. It was targeted for psychological impact, to deprive Noriega of his principal TV media, and to ensure prompt reception by the populace. To this end, and according to the OPLAN, Channel 2's broadcast capability was neutralized by special operations forces at approximately H-hour, and Volant Solo began broadcasting on the frequency. The broadcast facility was shut down in such a manner as to enable prompt restoration of capabilities once the US and new Panamanian government were in full control of the situation.

The PSYOP element commenced activities by making use of the same previously prepared scripts and recordings being broadcast via Volant Solo. Soon it was heavily involved in preparing scripts and acquiring news items and music adequate to keep the Panama City area receiving Volant Solo's broadcasts and the nationwide audience receiving the AM radio station's broadcasts 24 hours per day. This was a major undertaking: Scripts would have to be developed, coordinated with the staff, and translated into acceptable Spanish so the Panamanian population would listen and understand. All of the PSYOP soldiers were actively engaged in this process. Fortunately, all were bilingual and thoroughly familiar with the situation in Panama. Tapes of popular music were available to maintain audience attention between announcements.

Safe conduct passes were prepared at and distributed by the printing facilities at the main US military print plant at Corozal. (Later versions, as well as other products, were printed on presses deployed with the PSYOP TF and on captured presses at Noriega's former PSYOP facility on Fort Amador.) It should be noted that Gen Marc A. Cisneros, CG USARSO, signed the leaflet, as opposed to CINCSOUTH (General Thurman), or CJTF-South (General Stiner). General Cisneros had already established a good reputation among the Panamanian population and the PDF, and therefore was expected to have greater credibility as a known entity. As events proved, his reputation would be a valuable commodity, not only as endorser on subsequent surrender appeals, but in personally arranging the surrender of regional PDF garrisons by telephone.

These first safe conduct passes were printed on newsprint. Three hundred thousand of them could be printed and chopped into bundles rapidly in a form that could be handled by two of the cell's PSYOP sergeants with a pickup truck. Furthermore, US helicopter crews could easily take on 30,000 or more leaflets and toss them over target areas without difficulty due to their compactness and small size. If the

---

PASAPORTE A LA LIBERTAD

ESTE PASAPORTE ES PARA EL USO DE MIEMBROS DE LA F.F.D.D. BATALLON DE LA DIGNIDAD Y LA
COOEPADI. SI SE PRESENTA ESTE BOLETO DE LOS ESTADOS UNIDOS, LE GARANTIZAMOS SU
SEGURIDAD, ACCESO A FACILIDADES MEDICAS, COMIDA, Y UN LUGAR DE DESCANSO Y RECUPERACION.
RECUERDEN: NO HAY QUE SUFRIR MAS.

GENERAL MARC A. CISNEROS
COMANDANTE DE TROPAS DEL EJERCITO SUR

PASAPORTE A LA LIBERTAD

SAFE CONDUCT PASS

THIS PASS IS FOR USE BY PDF, DIGNITY BATTALION, AND COOEPADI MEMBERS.
THE BEARER OF THIS PASS, UPON PRESENTING IT TO ANY U.S. MILITARY MEMBER, WILL
BE GUARANTEED SAFE PASSAGE TO U.S. FACILITIES THAT WILL PROVIDE MEDICAL
ATTENTION, FOOD, AND SHELTER.

GENERAL MARC A. CISNEROS
CG, US ARMY SOUTH

SAFE CONDUCT PASS

Source: 1st PSYOP Battalion, US Army

Figure 2. Three hundred thousand of these safe conduct passes were printed on newsprint as an expedient. They were printed in Spanish on one side but in English on the other so that US troops understood what was being promised. These were proven effective in the early stages of *Just Cause,* especially when accompanied by loudspeaker broadcasts.

contingency had taken place during the rainy season, this approach might not have worked very well since the newsprint would have rapidly disintegrated on contact with water. Had heavier paper stock been used, however, the leaflets would have taken longer to print and might have been too bulky and heavy for the small detachment to handle.

These safe conduct passes were dropped by helicopter on targets of opportunity and on locations identified by PSYOP soldiers. Helicopter pilots often carried bundles of leaflets just in case an opportunity would present itself for their use. Later versions of these passes were also distributed as handouts from loudspeaker teams and other US forces.

One of the earliest indicators of effectiveness was the successful employment of loudspeakers and leaflets in support of US Marines at La Chorrera, a small village on the outskirts of Panama City and Howard AFB. The Marines had encountered heavy resistance from PDF and DIGBAT members for the first 24 hours. The Marines ceased fire for the night. Surrender appeals were then blared over loudspeakers as safe passage leaflets were dropped. Resistance ceased early the next morning without an additional shot being fired. Surrendering personnel came forward clenching the safe conduct passes.

Colocated with the PSYOP cell was the commander of the Joint Audiovisual Detachment (JAVDET), a unit comprised of multiservice elements from the Joint Combat Camera Center. Although not in command of the JAVDET, the PSYOP TF commander could request specific video coverage by the detachment. He then had a significant amount of footage already available when the main body of the PSYOP TF arrived with organic camera teams and audiovisual production capability. One copy of every cassette produced by the JAVDET was delivered to the PSYOP TF.

The JAVDET's transportation asset were extremely limited, but the PSYOP TF commander was able to match up these cameramen with several of the vehicle-mounted loudspeaker teams already in-country. It was largely the videos obtained by these photographers that were used in initial PSYOP audiovisual operations. Also, since these videos were not PSYOP products, they were provided by the JAVDET to the US media through program action officer (PAO) channels. PSYOP loudspeaker soldiers and activities therefore wound up receiving higher visibility in the US media than would have been the case otherwise.

Immediately upon activation of the 10,000-watt AM PSYOP radio station and commencement of the initial assaults on 20 December, the new president of Panama, Guillermo Endara, communicated with the Panamanian masses principally via this station. Among Endara's first announcements was the promise that anyone turning a weapon in to US forces would be paid $150 for it.

The "money for guns" concept had been envisioned as a viable PSYOP technique to promote cessation of hostilities. Results during *Urgent Fury* in Grenada had been successful, even though undertaken late in the operation. In the case of Panama, the initiative was taken early; it paid off. The PSYOP radio station explained and promoted turn-in procedures to combatants and to the Panamanian population. The newly arrived PSYOP TF commenced preparation of numerous printed leaflets and posters to convey information leading not only to the rapid surrender of PDF and DIGBAT personnel, but to their turning in individuals and cached weapons as well.

Early in the operation, PSYOP soldiers supported civil military operations, including the displaced persons (DP) camp hastily established in Balboa Village. A tactical PSYOP loud-speaker team set up full-time operations in the compound and amplified PSYOP radio broadcasts over the loudspeaker system to put the people at ease and keep them informed. This team, along with other PSYOP soldiers, moved throughout the camp, handing out newspapers prepared by the PSYOP TF, pretesting proposed PSYOP products, chatting with the people, observing their actions, and reporting potential problems to the camp administrators—members of the 96th Civil Affairs Battalion.

A primary problem associated with the DP activity was rumor control. By building rapport and establishing information programs with the DPs over time, PSYOP soldiers established their credibility and were able to neutralize rumors before they expanded to crisis proportion. When an improved, expanded DP facility was completed at the Panamanian-controlled portion of Albrook Field, PSYOP soldiers helped prepare the population of the Balboa camp to minimize apprehension about the move to the new site and to enhance control during the move itself.

The tempo of PSYOP increased dramatically as time progressed and the PSYOP TF became fully operational. Additionally, the 4th PSYOP Gp commander arrived as senior PSYOP commander in-theater—serving as special PSYOP advisor and expeditor for General Thurman. The group commander used the full range of media to convey public information while soliciting cooperation and support from the

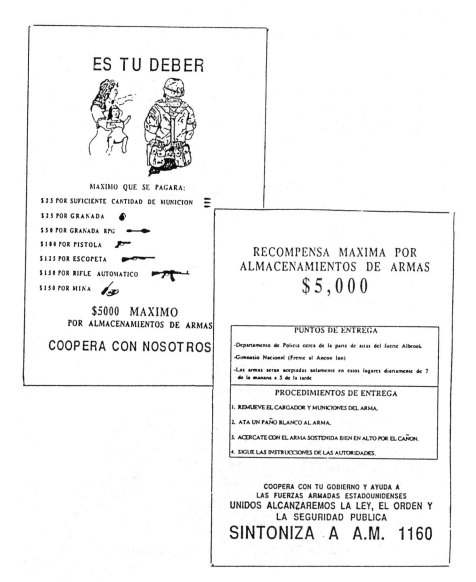

Source: 1st PSYOP Battalion, US Army

Figure 3. A series of posters and handbills advertised the reward program for weapons. (Interestingly, a Panamanian showed up in an armored vehicle and asked what the reward was for it. Since the maximum authorized amount for any one item was $150, US troops offered $150. After a short negotiation session, the Panamanian accepted the offer.)

Source: 1st PSYOP Battalion, US Army

Figure 4. Christmas cards were distributed during *Just Cause* to enhance relations and rebuild national spirit; 10,000 of the first and 30,000 of the second were distributed—mostly by hand and by helicopter drop.

military in identifying and apprehending PDF and DIGBAT members and arms caches.

Posters, leaflets, radio broadcasts, and television announcements solicited information and provided telephone numbers to be called by those wishing to volunteer information regarding weapons or PDF and DIGBAT fugitives. These techniques were effective. Within minutes after dissemination of printed material or broadcast of the information, the designated telephones began to ring and information poured in. It was also found that new posters, especially "wanted" posters bearing photographs of PDF and DIGBAT fugitives, had to be printed and distributed continuously because people removed them from walls to avoid identification, to keep as souvenirs, or possibly take them home for study and future reference.

Other products focused on establishing a favorable image of US forces, explaining US motives, and providing information needed to reestablish the normal routine. Products of the latter category were in demand because the former PDF-dominated Panamanian media was essentially shut down. In fact, the newspapers printed by the PSYOP TF were so sought after by the public that enterprising Panamanians grabbed bundles of them and sold them to passers-by.

At this point, it should be stressed that all products were tested prior to dissemination. Product review and approval was extended to the level of the executive office of the Panamanian government in order to reduce the probability of error. (A classic example of such an error on the part of an enemy occurred during *Desert Shield/Storm* when an Iraqi propaganda announcer warned US military personnel that their wives were fooling around with movie stars such as Bart Simpson!)

A short case study is presented here for the purposes of clearly demonstrating the complexity of PSYOP product development during *Just Cause* and, later, *Promote Liberty*. During the operation, the PSYOP TF developed a product bearing the multicolored image of the Panama national seal. The insignia was surmounted by an eagle. Due to the poster's size and complexity (five colors), the prototype was assembled and sent to the Corozal print plant. The workers at the plant

SE BUSCA

TTE CNEL LUIS A. CORDOBA

Ex-jefe oficial de inteligencia. Ex-miembro del Consejo Estratégico Militar CEM. Responsable por el arresto de los oficiales del G-2 envueltos en el golpe de marzo de 1988. Ex-jefe del DNTT y la 5ta zona militar. Trigueño, complexión mediana, cabello negro, ojos castaños, estatura 5'8' y peso 160 Lbs.

SI TIENES INFORMACION ACERCA DEL PARADERO DE ESTA PERSONA, LLAMA AL TELEFONO:
25-55-24

Source: 1st PSYOP Battalion, US Army

Figure 5. Several versions of these kinds of posters were employed by members of the Joint Task Force in attempts to round up former members of the Panama Defense Forces and Dignity Battalions. These posters, along with radio and television announcements, not only contributed to the prompt identification and capture of fugitives but also served to reassure the Panamanian populace that US forces were actively seeking to round up those individuals most capable of orchestrating retaliatory actions or restoring the "old order."

SE BUSCA

MAJ. CARLOS SALDAÑA
Fuerte complexión, cabello
negro, ojos castaños, rostro
ovalado.

HUMBERTO LOPEZ TIRONE
Sicario de linea dura. En-
lace con los paises socialistas y
lider paramilitar.

MAY GONZALO GONZALEZ
Promovido a Mayor luego del golpe
de Oct 3 por ayudar a rescatar a
Noriega. Se sospecha que ejecutó
varios oficiales que participaron.

ROLANDO ESTERLING
S-1 para el Batallón de
Dignidad en San Miguelito

CPT ASUNCION GAITAN

CPT BENJAMIN SOLIS

Llame a los teléfonos 87-3613, 87-4965 o 87-4246 si tiene información o fotografías de cualquiera de las personas aquí listadas. Llame también para reportar información de los batallones de dignidad o antigua Fuerza de Defensa.

Los siguientes individuos también son buscados por las autoridades de Panamá y EE.UU. Si tiene información de su paradero, o fotografías de ellos, por favor llame a los teléfonos 87-3613, 87-4965 o 87-4246

Eric Acsta	Jaime Jaen
Alberto Aleman Boyd	Guillermo Ledezma Bradley
Raman Argote	Donaldo Lopez
Camilo Aguilar	Ciro Macklean
Gumeriada Avecilla	Manuel Madarnas
Norman Baena	Victor Marucci
Natividad Barcenas	Alfonso Matthews
Angel Besiria	Faustino Diaz de McCrea
Laudegario De Bano	Sergio Morales
Romulo anchei Bettancourt	Carlos Moreno
Andres Buitrago	Alencar Ortiz
Manuel Carol	Ernesto A. Ortiz
Osvaldo Castro	Peters Renato
Eric Cedeño	Rennato Pereira
Nolvo Cadeño	Virgilio Periñan
Jorge Chandeck	Sergio Pittí
Orlando Cogley	Eric Polo
Mario Concepcion	Luis Peyaz
Alfonso Cordoba	Evidelio Quial
Santos Correa	Silvestre Rattaneer
Efigenio Cortez	Rhona E. Read
Victor Degraci	Fabio Rios
Gonzalo Delgado	Antonio Rodriguez
Jaime Le Dazman	Ernesto Rodriguez
Antonio Diaz	Guillermo Rosales
Mario Escudero	Rafael Saavedra
Felipe Escribi	Edilio Saldaña
Ramón Fernandez	Alga De Sanjur
Fonseca, Wilredo	Adriano Santos
José A. Ortega Gomez	Luis Chong Sing
Roberto Garrido	Hericlides H. Sucre
Luis Carlos Gomez	Omar Miranda Vega
Watson Gomez	Marcelle Tando de Ciacci
Gonzalo Gonzalez	A. Urquilla
Edgar Grimas	Romulo Chambonett Vasquez
Juan Hernandez	Carlos Antonio Whisroon
Teodoro Hunt	José Vernal
Roberto Iglesia	Andres Wkrago

Source: 1st PSYOP Battalion, US Army

Figure 6. "Wanted" handbill

The reproduced newspaper is largely illegible. Its main visible headings and text are transcribed below.

TRIBUNAL ELECTORAL LEGALIZA AL GOBIERNO DE ENDARA

Rodolfo Chiari de León, Ministro de Gobierno y Justicia bajo el gobierno militar del depuesto dictador M.A.N. dirigió una carta el 26 de Junio de 1989 a su excelencia Joao Baena Soares, Secretario General de la O.E.A., para garantizarle la constitucionalidad de la decisión tomada por el Tribunal Electoral al declarar nulas las elecciones generales celebradas el 7 de Mayo. Mediante este engaño pretendía la dictadura de Noriega tapar el cielo con la mano y legitimizar frente a la comunidad internacional el fraude criminal perpetrado al robarle las elecciones al pueblo panameño.

[remaining column text illegible]

LINEAS PARA REPORTAR INFORMACION

Llame para reportar información sobre almacenamientos de armas y municiones, actividades criminales de los difuntos batallones de indignidad, la antigua Fuerza de la Defensa, y la localización de los compinches de Noriega.

SINTONIZA A A.M. 1160

El Private Coons, del ejército de EE.UU. toma tiempo para jugar con algunos niños panameños. Un soldado comentó, "Estos niños son mas importantes para nosotros que Noriega."

CAPTURADO

COL. ANGEL ALBERTO MINA

Con tu cooperación los capturarémos a todos. Llama a los teléfonos 287-6453 o 287-5965.

CON CAUTELA DISCUTEN LOS RUMANOS EL FUTURO

TIMISOARA, RUMANIA (UPI) - [remaining column text illegible]

Source: 1st PSYOP Battalion, US Army

Figure 7. Note that this PSYOP newspaper advertises the radio station frequency, announces a recent capture of a PDF officer, and provides telephone numbers for reporting locations of arms, munitions, criminals, and criminal activities. Panamanians picked up bundles of these newspapers and sold them.

remembered that they had produced similar images before the contingency and still had the master plates; they therefore used these plates. The resulting product was indeed beautifully done. Only one problem: the PSYOP TF S-3 realized that, instead of the traditional eagle atop the Panama crest, there was a North American bald eagle—not exactly the kind of thing that would make us popular!

When redone, the revised poster with correct eagle and crest, was tested again "just to be sure." Again, a problem surfaced—according to the National Code of Panama, the crest was supposed to appear in a field of green, and the proposed poster would have done just that. Unfortunately, the proposed green was the color of one party of the Panama coalition government—one which was becoming increasingly viewed by Panama political opponents as trying to muscle out the other two coalition parties.

The poster finally came out with the eagle and crest on a neutral background. The lesson here is that pretesting and posttesting, including the actual colors to be used, are fundamental to the PSYOP development and production process.

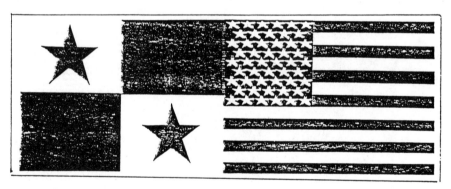

Source: 1st PSYOP Battalion, US Army

Figure 8. This small red, white, and blue poster was printed on heavy stock paper and was distributed throughout _Just Cause_ and _Promote Liberty_. It was displayed on military vehicles and handed out by US troops. Its objectives were to portray US-Panamanian teamwork and to reduce perceptions of US forces as military invaders or occupiers. Within days after dissemination commenced, Panamanian commercial vendors were peddling T-shirts and other paraphernalia bearing the same logo.

This process is crucial—and testing needs to be done right. Additionally, all PSYOP campaigns and products require extensive staff coordination and personal review by the PSYOP TF commander.

As the situation further stabilized, it became increasingly important to get the commercial media "back in business." Shortly after the first day of operations, the PSYOP TF commander began to receive personal telephone calls from media representatives willing to get their TV and radio stations back on the air in support of the US and the new government— but only if they and their personnel would be escorted by US soldiers or transported under guard while their facilities were guarded by troops of the JTF. There existed considerable concern for possible terrorist acts by DIGBAT goons still at large, including possible murder or sabotage.

Initially, soldiers and vehicles could not be spared by either the PSYOP TF or JTF-South to satisfy all commercial media's conditions. Nor did the PSYOP TF have the weapons necessary to provide appropriate security. However, at about 2100 hours the night after Noriega obtained sanctuary in the Papal Nuniciature, a team of PSYOP soldiers entered the TV Channel 2 broadcast facility and began broadcasting prerecorded tapes of cartoons from the station library. By morning, most of the commercial broadcasting crew had arrived, along with a large crowd of Panamanians enthusiastically waving Panamanian and US flags. They were celebrating the arrival of US troops and the "grand reopening" of their singing channel.

The commercial crews volunteered to restore operations as the PSYOP team temporarily observed and screened their material to ensure that it was not counterproductive to US aims or those of the new government. Another TV channel (4) was similarly reactivated the same day.

Most of the principal radio and television stations in and around Panama City were back on the air within days, temporarily under observation by or supervision of either US PSYOP personnel or Panamanian officials with whom PSYOP had established a working relationship. These media not only broadcast materials prepared by the PSYOP TF, but also prepared and broadcast excellent programs on their own. Their programs supported US military forces and the government of Panama.

By 8 January, the PSYOP TF had produced and disseminated over one million leaflets and handbills, 50,000 posters, 550,000 newspapers, and 125,000 units of miscellaneous other printed materials. Volant Solo had conducted TV broadcasts for the first several days, the PSYOP AM radio station had been operating 24 hours a day, and countless messages had been aired on commercial radio and television stations and published in commercial newspapers. Loudspeaker teams continued to support tactical units by broadcasting advisories, interacting with the population, and providing timely PSYOP advice to US commanders.

Psychological operations figured prominently in the successful US efforts to flush Noriega out of the Papal Nunciature, where he had sought refuge. They played a significant role in establishing the Panama Public Force—the new police force designed to replace the old PDF. PSYOP teams circulated through Panama City and other population areas, assessing conditions and public attitudes, interacting directly with the public and with US combat and security forces, and resolving difficulties when they arose.

All of the first series of US PSYOP objectives were accomplished during *Just Cause*. US combat elements began to depart in early January 1990. Focus shifted to the consolidation of the new government in Panama and the restoration of the public information infrastructure. The PSYOP TF was stood down, replaced by a PSYOP support element comprised of 48 personnel.

This element, consisting of stay-behind PSYOP soldiers from *Just Cause* and a civilian analyst from the 1st PSYOP Bn, provided PSYOP support to the newly created Military Support Group. This unit then provided US military support to the government of Panama during the nation assistance and civil-military operations of *Promote Liberty*. From mid-January to the end of February, the PSYOP support element produced and disseminated more products and planned and executed more activities than all those in support of *Just Cause* combat operations.

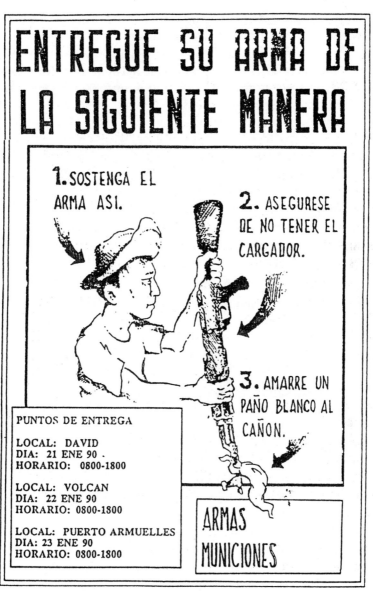

Source: 1st PSYOP Battalion, US Army

Figure 9. One of many later versions, this poster not only indicates procedures for surrendering arms but also identifies turn-in locations and their hours of operation. Subsequent versions left some of the data blank so that Special Forces soldiers in the countryside could fill in the information and administer similar programs.

The highly efficient and productive mini-task force increasingly worked in conjunction with the new government, the commercial media, the US Embassy, the Military Support Group, and JTF-Panama to create conditions favorable for restoring democracy, the economy, and public security. Major emphasis was placed on these objectives:

- Maintaining support for US objectives in Panama.
- Professionalizing the newly formed Panama Public Force and Panamanian Police Force (PNP)—constituted principally by former members of the PDF.
- Gaining public support for the Public Force and PNP.
- Enhancing the effectiveness and prestige of the new government.
- Neutralizing disinformation and hostile propaganda directed against the government, the Public Force/PNP, and the US.

By the end of February, these PSYOP soldiers had been redeployed to Fort Bragg. They were replaced by 8th PSYOP Bn soldiers augmented by linguists and area experts from the 1st PSYOP Bn and by camera and audiovisual technician teams from the 4th PSYOP Gp's Strategic Dissemination Company. This group continued operations, employing the full range of dissemination media, until stood down in May 1991. The final PSYOP objectives envisioned for Panama contingency operations had been accomplished.

From beginning to end, PSYOP supported all stated US military and political objectives. During the initial stages of the assault, PSYOP loudspeakers, leaflets, TV and radio announcements successfully appealed for cessation of hostilities; surrender of weapons and members of the PDF and Dignity Battalions; and compliance with measures promulgated by the newly formed Panamanian government. Psychological operations enhanced credibility of US military forces, assisted in gaining and retaining popular support for US actions, and helped reduce the potential for damaging disinformation and hostile propaganda.

As the operation progressed, psychological operations and PSYOP forces were fully integrated into consolidation operations and emergency relief activities while fostering an

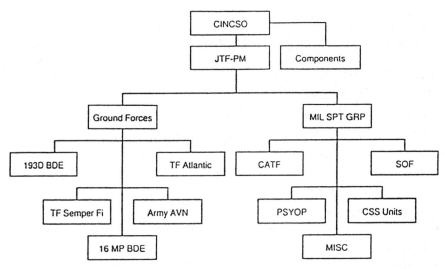

Source: 1st PSYOP Battalion, US Army

Figure 10. The PSYOP support element coordinated directly with JTF-PM, ground forces, other elements within the MSG, the Panamanian government, and the US Embassy.

atmosphere of return to normalcy. Announcements via US military PSYOP and commercial news media supported this process. Through programs and information developed or disseminated by PSYOP soldiers, 193d SOG Volant Solo technicians, and host-nation affiliates, Panamanians knew when and where to go to satisfy their fundamental needs. Through this process, PSYOP supported not only the theme of legitimacy of the US operation, but also the legitimacy of the newly formed government of Panama.

During the final phase of military operations, US PSYOP supported an orderly return of control to the host-nation government, addressed major sanitation and health threats, and continued military information programs. All of this facilitated prompt return of the deployed US military personnel to CONUS.

Finally, as a natural consequence of these accomplishments, psychological operations and PSYOP units gained recognition among many important US commanders of combat and combat support units participating in the operation. Not only did PSYOP affect them personally by assisting in the reduction of casualties

273

and damage on both sides, but also assisting in presenting the most favorable image of their commands to the Panamanian population and to the world.

Many of these accomplishments cannot be measured in quantifiable terms. However, there were obvious indicators of success:

- surrender of enemy forces after exposure to loudspeakers and leaflets;
- immediate public response to posters, loudspeakers, and TV/radio announcements, resulting in valuable information and intelligence;
- civilian compliance with instructions conveyed via PSYOP media;
- virtual absence of hostile propaganda and disinformation which could have obstructed the mission; and
- prompt restoration of commercial information services, which contributed to the legitimacy of the new democratic government.

Many valuable lessons were learned by the PSYOP community itself:

- The PSYOP TF commander simultaneously serving as JTF PSYOP officer, theater PSYOP officer, and PSYOP TF commander enhanced initial planning and coordination. As the operation progressed, however, these duties overextended the capabilities of the commander to the extent that he encountered difficulty supervising and controlling the many disparate and geographically separated PSYOP elements and operations.
- The requirement that PSYOP TF simultaneously provide JTF staff support, control tactical assets, and develop and disseminate PSYOP products stressed the capabilities of the task force to the breaking point. The battalion staff and company configurations did not easily adjust to the requirements of a task force much larger and more complex than experienced before. Of particular concern was an inability to control the tactical PSYOP teams, which were vital to successful PSYOP campaign planning and evaluation. Attaching these teams OPCON to combat commanders while optimizing on-site responsiveness during the assault phase detracted from

overall PSYOP TF performance during consolidation activities. These teams should have reverted to a direct support status under operational command of the PSYOP TF as soon as that force had the ability to fully exercise that function. Unfortunately, that capability did not materialize until the majority of the teams had returned to the US.

- Augmentation from other battalions of the 4th PSYOP Gp was essential to the success of the PSYOP TF. However, due to the peacetime configuration and functioning of the group, which only occasionally involved an integration of effort across battalion lines, the full talents and capabilities of soldiers and equipment were extremely difficult to exploit in Panama.

- The L-series modified table of organization and equipment (MTOE) for the 4th PSYOP Gp and its subordinate units was not optimally suited to requirements of the contingency.

Some of the "good things" that were either learned or reinforced are listed below.

- Early integration of PSYOP into the contingency planning process helped ensure overall success of the operation.

- Tactical loudspeaker teams contributed to demoralization of the PDF and DIGBATs and to cessation of hostilities and surrender. The teams used prepackaged scripts and recordings, but frequently employed their own ingenuity and linguistic capability "on the spot" to deal with both hostile forces and the civilian population. Tactical commanders of Army forces, Marine Corps forces, and Navy SEAL forces placed considerable importance on these teams.

- The Panama population listened to, and complied with, instructions and advisories from US military PSYOP TV, AM radio, and loudspeaker broadcasts.

- The working relationship that the 4th PSYOP Gp commander and the 1st PSYOP Bn commander enjoyed with the XVIII Airborne Corps commander and staff, and with CINCSOUTH and his staff prior to Just Cause and Promote Liberty, undoubtedly resulted in a degree of understanding and trust which were critical to the success of joint PSYOP throughout the contingency operation.

- Psychological operations soldiers and airmen were well trained, motivated, self-confident, and effective.

Certainly, some of the problems and some of the "success stories" were situation dependent and can never really be formalized. However, in June 1991—less than six months after the Panama contingency was executed and only weeks after return to CONUS by the last 8th PSYOP Bn members of the PSYOP Support Element—the 4th PSYOP Gp was provisionally reorganized. Regionally oriented battalions remained, but the group was now organized according to an operational concept suitable for a wide variety of contingencies.

The reorganized 4th's basic tenets:

• The 4th PSYOP Gp commander would deploy to the theater headquarters, along with a small headquarters element, to provide PSYOP advice and assistance to the supported theater CINC and act as "expeditor" for PSYOP-related actions. (This happened informally in Panama, when the 4th PSYOP Gp commander performed this function not only for the PSYOP TF but also for the 96th Civil Affairs Bn, the 112th Signal Bn, and the 528th Support Bn—three units which were at that time under the 4th's peacetime command and control. The concept proved so beneficial to both the PSYOP TF and to CINCSOUTH that it was formalized.)

• The PSYOP TF would continue to be commanded by the theater PSYOP battalion commander, but with two other PSYOP battalion commanders working for him:

1. A tactical commander (the commander of the reconfigured 9th PSYOP Bn) would control all PSYOP loudspeaker teams and other tactical PSYOP assets and ensure support to other tactical elements of the JTF. The 9th PSYOP Bn would train for these support missions worldwide.

2. A media production commander (the commander of the new provisional PSYOP Dissemination Bn) would produce all printed products, recordings, and audiovisual products, and would run all radio and television broadcast operations.

Since the 4th PSYOP Gp would now be provisionally organized to operate the same in peacetime as in war, it was presumed that overall effectiveness of contingency PSYOP would improve.

The opportunity to validate the new concept presented itself sooner than anyone could have expected. Within one month after the 4th PSYOP Gp's reorganization, Iraqi forces invaded Kuwait. When PSYOP forces deployed to Saudi Arabia, they did so in the configuration described above. It worked, and worked well! It is estimated that 29 million leaflets and many hours of loudspeaker and radio broadcasts During *Desert Shield* and *Desert Storm* resulted in thousands of Iraqi soldiers being influenced by PSYOP to abandon their positions and equipment and to surrender.

One can only speculate on what might have been the results had PSYOP forces not participated in these contingencies, but the indicators of success are pretty clear. In the author's opinion, these successes are a tribute to the PSYOP soldiers, civilians, and airmen—past and present—whose competence, dedication, and ingenuity most certainly saved US and Panamanian lives and contributed to mission success. Their accomplishments have resulted in increased recognition of the value of PSYOP by senior military commanders and their staffs, and by other agencies of the US government. However, this renewed interest and understanding is highly perishable. We still have a long way to go before the potential of strategic, operational, and tactical PSYOP—and the forces which plan and execute these operations—is fully realized.

Psychological Operations in Action

Poland's Underground Media

Laurence J. Orzell

Poland offers a fascinating example of nongovernmental psychological operations in action. Drawing upon that country's long tradition of a conspiratorial press, dissidents affiliated with the banned Solidarity trade union and other organizations engaged in the large-scale production of underground newsletters, leaflets, books, and other printed materials designed to influence popular perceptions. The dissidents also utilized electronic media such as radio and cassettes.

Those activities reflected a broad range of antiregime opinion, but their overall purpose was twofold: to supply uncensored information on political, social, and economic issues, and to maintain a spirit of opposition among the populace. While the dissidents would probably have described their efforts as purely informational or journalistic, their activities could accurately be described as peacetime, nongovernmental PSYOP. This article places the Polish opposition's PSYOP efforts within their historical context but focuses on the period following the imposition of martial law (December 1981). In particular, the analysis examines the type of underground print materials produced, the means used to prepare and disseminate them, the government's response, the main themes the dissidents sought to propagate, and the opposition's use of electronic media. The article concludes with an assessment of the overall significance of those efforts.

The Polish resistance, whether directed against czarist Russia, Nazi Germany, or native communist rulers, recognized the value of PSYOP. When the Poles revolted against Russia in January 1863, they formed a conspiratorial government that not only conducted urban guerrilla warfare; it also organized an underground press which contained manifestos directed at

Poles, Lithuanians, Byelorussians, and Ukrainians. Although this effort failed, it contributed to the maintenance of armed resistance. The underground media played a much greater role during World War II. The Home Army *(Armia Krahiwa)*, Poland's chief resistance group, regularly published an *Informational Bulletin (Biuletyn Informacyjny)*, copies of which circulated from hand to hand and thus indirectly supported diversions and armed opposition to the Nazis. Other groups, ranging from the far right to the extreme left on the political spectrum, also produced leaflets *(ulotki)* and other publications. So extensive was this network that the London-based Polish government-in-exile issued a postage stamp in 1943—itself a form of PSYOP—commemorating the underground press *(prasa podziemna)*.

Armed resistance to foreign domination proved impractical following the communist consolidation of power after World War II, but dissident psychological operations eventually came to assume even greater importance. Antiregime intellectuals benefited from their government's reluctance to engage in extensive repression during the 1970s—lest Poland risk losing the economic benefits of détente—and they established a variety of illegal publishing ventures. One group in particular, the Workers' Defense Committee (later renamed the Social Self-defense Committee), distinguished itself through its publications: *Communique (Lomunikat)*, *Information Bulletin (Biuletyn Informacyjny)*, and *The Worker (Robotnik)*. Dissidents also set up an Independent Publishing House, better known by its Polish acronym NOWA, which produced numerous literary works and several periodicals. Despite periodic harassment by the authorities, NOWA and similar illegal publishing enterprises prepared materials targeted at laborers, students, farmers, and other social groups. There can be little doubt that these publications, which advocated political pluralism and respect for human rights, helped forge the broad social alliance which led to the formation of Solidarity during July and August 1980.

Both the regime and the workers used PSYOP extensively during the 1980 strikes. For instance, when the authorities distributed leaflets questioning the striker's motives, the workers responded with broadsheets of their own, counseling resistance in the face of what they termed government disinformation. The government's acceptance of the strikers' "Twenty-one Demands" effectively extended legal recognition to unofficial publishing activities. The subsequent "Gdansk Agreement" obliged the regime "to respect freedom of speech and . . . not to suppress independent publications." Solidarity quickly took advantage of this unprecedented concession. The union published its own national organ, *Solidarity Weekly (Tygodnik Solidarnosci)*, and regional branches followed suit with a plethora of journals and newsletters. The union press both reflected and shaped the popular perception that the government could not be trusted to carry out the Gdansk accord. Not surprisingly, these publications represented one of the principal targets of repression when the authorities declared martial law on 13 December 1981.

Despite internment, arrests, and confiscations, the underground press quickly reappeared and assumed an even greater importance due to the fact of dissatisfaction. Some activists who were placed in internment camps in 1981–82 even managed to produce handwritten newsletters. For instance, internees at a detention camp in Nowy Lupkow prepared several issues of a newsletter entitled *Our Bars (Nasza Krata)*, named after the prison bars which isolated them from society. Copies of *Our Bars*, reportedly circulated outside the camp, counseled the maintenance of worker-intellectual ties in the face of government repression. The vast majority of underground publications produced during and after martial law emanated from small groups of individuals in larger cities and towns who had avoided capture. The importance of these PSYOP efforts in dissident strategy became clear when Solidarity's underground Provisional Coordinating Committee issued a declaration in 1982, stressing the need to "organize the independent circulation of information" and thereby "expose the propaganda goals of the authorities."

The underground press produced nearly 1,000 different periodicals, approximately 400 of which appeared repeatedly, some on a regular basis. While it is difficult to estimate the total readership, the sheer number of publications suggests that these materials enjoyed wide circulation, primarily in urban areas. During the first four months of martial law, the underground published 100 different titles. Many of the early editions bore the simple yet historically significant title *Information Bulletin (Biuletyn Informacyjny)* and consisted of one or two pages. As might be expected, the majority of these materials were prepared in Warsaw, Poznan, Wroclaw, Lodz, and Gdansk. Press runs varied from several hundred to over 10,000, depending largely upon available equipment and printing supplies. During the martial law period (December 1981–July 1983), most of these publications emanated from the remnants of these periodicals, including *Mazovian Weekly (Tygodnik Mazowsze)*, which was prepared by union activists in Warsaw; but Solidarity cells in factories and other local enterprises also issued short newsletters.

Solidarity activists did not, however, constitute the only source of clandestine periodicals in Poland. The publications of other underground groups assumed somewhat greater prominence after 1985, due at least in part to the increasingly dim prospects for the union's revival. Nearly all of these groups professed loyalty to Solidarity's ideals, but they also expressed a variety of opinions that distinguish them from the Solidarity mainstream. Three of these organizations merit particular attention. One, known as the Committees for Social Defense *(Komitety Obrony Spolecznej)*, issued a journal called *KOS*, which was directed primarily toward an intellectual audience. The Fighting Solidarity *(Solidarnosc Walczaca)* group published a paper of the same name, advocating a more confrontational stance than that favored by most Solidarity leaders. The Confederation for an Independent Poland *(Konfederacja Polski Niepodleglej)*, a right-of-center group that antedated Solidarity, disseminated its strong critiques of the regime through its journals *Independence (Niepodleglosc)* and *We Don't Want Commies (Nie chcemy komuny)*. Further, publications targeted at specific social groups such as teachers, university students, and medical workers also attained greater prominence.

Several clandestine publishers branched out into the production of full-length books, an effort that gave rise to greater coordination among the underground media. In October 1986, representatives of several underground journals formed a Social Council for Independent Publishers. At the same time, several publishers concentrated on books rather than periodicals because sales of the latter proved less profitable. Because this might portend a shift from shorter periodicals to longer products, some activists called for renewed emphasis on journals. Much of the debate centered on the role of the Consortium of Independent Publishers, a group founded in 1986, which sought to promote greater cooperation and minimize duplication within the underground media. Defenders of the consortium claimed that periodicals would continue to play a key role. In June 1986, the consortium helped found an Independent Publishers' Insurance Fund to reimburse members whose operations were closed by the authorities.

Leaflets, handbills, and other ephemera were also used by the Polish opposition, but they declined significantly with the decline in large-scale demonstrations. During the martial law period, the underground produced leaflets in connection with local strikes and other public protests. For example, activists printed and distributed a small leaflet announcing the place and time of a pro-Solidarity rally in Warsaw scheduled for May Day 1985. Leaflets were also used to alert the populace of the time and frequency of Radio Solidarity broadcasts. Still other broadsheets were designed to identify and disgrace those who cooperated with the regime during martial law. For instance, one handbill, entitled "List of Collaborators," purported to be a commendation from the government. The citation praised the recipients for their assistance and contained a blank space for insertion of the collaborators names.

Polish underground PSYOP also relied heavily upon the dissemination of nonverbal symbols by means of mock postage stamps and currency. In 1984, for example, the opposition produced a series of stamps portraying dissident leaders such as Jacek Kuron and Adam Michnik. That same year, the Solidarity organization in Krakow issued a stamp in honor of George Orwell. During 1986, dissidents printed a set of stamps commemorating "150 Years of the Underground Press in Poland." In addition, the underground produced forged Polish currency bearing the portrait of Lech Walesa in place of the figures who appeared on genuine banknotes. Dissidents reportedly printed a banknote containing the likeness of Gen Wojciech Jaruzelski and the inscription, "30 Pieces of Silver." Crude forgeries of US currency, designed as novelty items, also appeared. During 1985, for instance, an oversized counterfeit of a $100 bill appeared in Poland with Walesa's picture in place of Benjamin Franklin's portrait.

The producers of underground publications faced a variety of obstacles. Dissidents understandably were reluctant to discuss openly the details of production and dissemination techniques. Most individual publishing enterprises were comprised of small numbers of people who strived to keep their identities and the location of their equipment well-concealed. Materials were first produced on typewriters using carbon paper or on mimeograph machines but the opposition increasingly came to rely on offset and silk-screen presses. The printing process was itself dispersed, particular functions being assigned to different locations. According to the production chief, Warsaw's *Tygodnik Mazowsze* needed 37 apartments, 30 of which were used in preparing each issue. Funds to acquire and maintain equipment came from publication sales, but Solidarity sympathizers abroad also collected financial assistance and made it available. The dissemination of underground publications necessarily involved significantly large numbers of people and great risk. Once printed, copies were reportedly sent to a variety of locations, where *kolportery* (literally, "hawkers") collected them for delivery to subscribers. In the case of *Tygodnik Mazowsze*, this led to the creation of a huge distribution network.

The Polish government responded to underground publishing with a curious strategy that blended harassment, repression, propaganda attacks, disinformation, and apparent neglect. The regime attempted to limit underground publications through strictures on the availability of printing supplies. Stationery shops were required to record all purchases of more than 1,000 sheets of paper. Agents of the Security Service *(Sluzba Bezpieczenstwa)* reportedly monitored such purchases and penetrated several underground organizations. From time to time, officials arrested individuals connected with dissident publications. Anyone engaged in such activity could be detained for 48 hours and fined up to 50,000 zlotys—the amount earned by an average Pole in two months. The authorities sometimes confiscated automobiles belonging to distributors, presumably to hamper the latter's ability to continue their work. In some cases, the Security Service closed printing operations completely.

The regime also conducted a major PSYOP effort of its own to discredit the underground media. This campaign contained a variety of themes, one of which posited that underground publications were tools of the West—especially the United States. For example, the official press frequently claimed that US dollars were sent to the underground to ferment unrest in Poland. Also, infrequent factual errors in the underground media offered opportunities for the regime to attack their credibility. Moreover, government agents periodically engaged in black propaganda. In some instances, portions of genuine publications were excised and replaced with forged texts. On other occasions, security officials produced forgeries of entire journals. And in still other cases, erroneous or otherwise embarrassing information was covertly supplied to publishers who reproduced it in good faith. For the most part, however, underground publishers discovered and exposed such activities.

Despite governmental attempts to close them down, Poland's underground media remained prodigious by East European standards. True, the sheer extent of underground publishing rendered it difficult for the regime to suppress such activity entirely, even if Polish officials had made a concerted effort to do so. Nevertheless, why they did not adopt even stronger measures raises some interesting questions. It may well be true, as one Western journalist suggested, that the government could not stamp out the underground media without causing uproar at home and making Poland an international outcast. This consideration might have grown even greater after Mikhail Gorbachev assumed power in the USSR. The regime might also have concluded that it could not hope to win genuine popular support and must therefore tolerate the phenomenon. Then, too, government officials might have believed that the underground press allowed dissidents to vent their frustrations in a relatively harmless manner so long as opposition remained largely confined to the realm of theory.

The underground press criticized and challenged the political, social, and economic status quo. During the martial law period, dissident media served primarily to report instances of repression and to encourage resistance among the populace. For example, they published information on political prisoners, endorsed strikes and demonstrations, and advocated noncooperation with the regime (e.g., election boycotts). When it became apparent that direct confrontation would not bring about the restoration of Solidarity (or other reforms), the underground press as a whole resigned itself to a long-term struggle with the authorities.

Underground publishing reflected a variety of political trends, from right-wing Polish conservatism to liberal democratic socialism. However, nearly all publications shared some common goals. According to a report issued in early 1987 by the Social Council for Independent Publishers, they strived to propagate undistorted Polish history, Poland's struggles for independence, the postwar rule of communism in Poland, and methods of national resistance. In this way, they hoped to expose "ideological myths" and "propaganda falsehoods of the communists."

In its attempts to achieve these goals, the underground press addressed a wide range of domestic and international issues. Violations of human rights and the absence of genuine democracy in Poland continued to bulk large in its reportage, but it also devoted significant attention to economic affairs in light of the chronic shortages and debt problems that plagued the Poles. It reflected and encouraged debates over the extent to which dissidents might work for reform through participation in legal organizations such as the workers' councils that were set up following the banning of Solidarity.

Significantly, the emergence of the Freedom and Peace Movement (Wolnosc i Pokoj), which encouraged arms reduction and resistance to compulsory service in the Polish People's Army, also received favorable attention from the underground media, as did environmental issues. Particularly noteworthy was the underground media's increased coverage of international questions such as US-Soviet arms negotiations, Gorbachev's apparent efforts at internal reform, and economic developments in China. Despite glasnost and perestroika, however, the underground media generally continued to criticize the Soviets and praise the Americans.

The underground press clearly constituted the principal means for the dissemination of opposition PSYOP, but dissidents also extended their efforts to the realm of electronic media. Radio broadcasting proved to be highly vulnerable to jamming and to detection by the authorities, but it continued to resurface from time to time. Dissidents with some experience in electronics constructed ersatz transmitters and made brief broadcasts on FM frequencies from the roofs of large buildings. Collectively called Radio Solidarity, such transmissions first appeared on the airwaves during April 1982, after which they were broadcast in several different cities. Program content was necessarily limited, but broadcasts usually addressed such current affairs as price hikes or the significance of historical anniversaries. In some instances, these transmissions successfully overrode television signals in order to convey a brief verbal message or a "still" image of the slogan *Solidarnosc zyje* (Solidarity lives).

Dissident psyopsers also made good use of videocassette recorders (VCR), audiotape recorders, and even microcomputers. Poland contained an estimated 500,000 to 700,000 privately owned VCRs. The NOWA underground publishing enterprise distributed not only Western films but also videocassettes on Polish subjects produced by the Paris-based Videoknotakt organization. Domestic productions, such as filmed interviews with Lech Walesa, also circulated widely. Government authorities attempted to suppress this activity—primarily by confiscations at the border—but the trade continued to flourish. The underground also prepared and disseminated a large variety of audiocassettes containing antiregime songs, news commentaries, interviews, and lectures on Polish history. Some of these tapes reportedly were broadcast over factory loudspeaker systems. Finally, several clever computer owners managed to generate antiregime texts or games. According to an acerbic report published in the hard-line military newspaper *Zolnierz Wolnosci (Soldier of Freedom)*, Western interests inspired some programmers to create programs that promoted anticommunist attitudes.

Despite various crackdowns by the authorities, the underground media remained a part of the Polish sociopolitical scene. Dissidents stood in a centuries-old tradition of conspiratorial information activity, and their eagerness to utilize modern technology rendered it virtually impossible for the regime to suppress their efforts. As suggested above, the government might have decided to tolerate this situation; nevertheless, the opposition media in Poland provided a real-world example of the success that can be attained through nongovernmental PSYOP.

The Libyan Raid as a Psychological Operation

Col Frank L. Goldstein, USAF

The use and employment of US forces in psychological operations is a basic but little understood part of US doctrine. The use of US forces to achieve national strategic and tactical security objectives is the basis of PSYOP planning. Security objectives include the influencing of friendly, neutral, and/or enemy audiences to behave favorably or unobstructively toward US national security. A psychological operation is any operation that conveys selected information and indicators to foreign individuals and groups and influences emotions, motives, objective reasoning, and ultimately the behavior of the target audience. Such an operation should also induce or reinforce foreign attitudes and behavior that are favorable to US objectives. The target audience is a foreign group that may include hostile military forces but can also be neutral or friendly.

The stated purposes of the US raid on Libya as reported in the press were to emphasize the cost of state-sponsored terrorism, to damage terrorist operations, and to encourage internal insurrection (*New York Times*, 5 May 1986). I view the US Libyan raid as strictly a psychological operation, an operation planned and carried out to achieve a psychological point of view.

An analysis of the Libyan situation prior to the US raid would have revealed the following scenario.

• Was a psychological operation required? The answer would be yes. The United States needed to influence attitudes and behaviors concerning terrorism among enemy, friendly, and neutral audiences, especially with Libya and Muammar Qadhafi.

• What would be the best psychological operation approach? The best approach in consideration of prior attempts to influence Qadhafi and others would be a limited tactical air strike.

• What was the psychological situation? The psychological situation revealed that many nations, friendly, neutral, and enemy, were not taking seriously the US concern about state-sponsored terrorism. Colonel Qadhafi had revealed himself to have mood swings and to be susceptible to depressive episodes. Intelligence and the media revealed that there was underlying dissatisfaction within the Libyan government with the current policies of Qadhafi. Additionally, the world had just been subjected to a series of terrorist acts and appeared receptive to such a raid.

• Would an air strike have a strategic psychological effect? A successful air strike against Libya and Qadhafi could produce the following: (a) notice to all nations that the US would not tolerate state-sponsored terrorism, (b) notice to Qadhafi that his power was not absolute, (c) notice to anti-Qadhafi forces in Libya that the current actions of Libya could reap dire consequences for all, (d) a reduction in terrorist acts as Qadhafi and his forces would be forced to regroup, and (e) additional time to get other Western nations more involved in antiterrorist operations. On 15 April 1986, the United States, using Air Force and Navy resources, carried out a bombing raid on Libyan targets.

The results of that raid among enemies:

• Qadhafi instructed his operatives to deemphasize attacks on US military targets but to look for easier US targets (*Washington Post*, 4 May 1986).

• The Soviets contended that Libya's failed defense was a failure of men, not weapons—some Libyan officers were purged.

• The absence of any promised Soviet military support and the fact that the Soviet Union and Libya did not negotiate a formal treaty pledging Moscow to come to Colonel Qadhafi's aid demonstrates that the USSR was not willing to back up Qadhafi with more than words (*Christian Science Monitor*, 24 April 1986).

• There was a decline in Qadhafi's popularity—and some decline was already ongoing due to economic policy.

• The Soviets openly questioned Qadhafi's wisdom—a high Soviet official was quoted as saying "Khadafy is a madman on top of a pile of gold" (*Washington Post*, 24 April 1986).

• Syria publicly and openly rejected the idea that Syria has any connection with terrorist activities.

• The Soviets found themselves in a double bind: if they helped Qadhafi too much they would drive him further away from conservative Arab support. If they didn't help him, he might destroy himself and the Soviet foothold in the Middle East.

Results of the raid among neutral nations:

• Sixty-five percent of the French people supported the United States and were critical of their own government's nonsupport (*Washington Post*, 24 April 1986).

• Saudi Arabia rejected a Libyan appeal for financial aid (*Washington Post*, 1 May 1986).

• Several Arab nations, while delivering perfunctory denunciations of the United States, privately expressed congratulations. Two Arab countries didn't even issue any complaints (Iraq and Tunisia). Egypt's, Oman's, and United Arab Emirates' official responses were noticeably mild. The newspapers of Jordan and Tunisia did not run editorials on the action (*Los Angeles Times*, 27 April 1986).

• A Libyan call for an emergency Arab summit meeting resulted in Iraq, Jordan, Syria, and Saudi Arabia stating they would be unable to attend.

The raid's results among friends:

• There was strong support against terrorism at the Tokyo Economic Summit.

• European allies were persuaded to take basic concerted action against terrorism—a blow for reason as well as deterrence (*New York Times*, 24 April 1986).

• Despite an initial outcry, west European countries moved with quiet determination on four fronts: They kicked out a number of Libyan diplomats and suspected agents, significantly increased intelligence through sharing and coordination, encouraged their nationals to leave Libya, and quietly cut much of their trade with Libya (*Boston Globe*, 25 April 1986).

In terms of Colonel Qadhafi's own personality, he made no major radio, TV, or personal appearance for almost six months. On the 16th anniversary of his expelling the US from Wheelus Air Base, a major event in Libya to which the foreign press was invited to hear a major speech, Qadhafi did not appear. Fewer than 2,000 attended. Videotapes were presented, revealing an exhausted-looking Qadhafi fidgeting in his chair, speaking in a hoarse voice, and making a speech devoid of his usual fiery rhetoric. This lackluster performance reinforced the belief that Qadhafi was slow to recover from the shock of the 15 April raid (*Time Magazine*, 23 June 1986). He began to reappear late in August 1986 in both public and TV formats.

The raid's effect on underlying dissatisfaction within Libya can be supported by the following comments:

- Qadhafi is still in control but soul-searching is under way. He began relooking at Libyan efforts in Chad and Sudan. Some senior Libyan leaders have talked openly about a lower Libyan profile in terrorism (*Washington Post*, 4 May 1986).
- The US raid forced Qadhafi to share power with the four members of Libya's ruling Revolutionary Council (*Time Magazine*, 23 June 1986).
- A summary of several stories can equate to the notion that the raid may have strengthened Qadhafi in the limited sense that the masses may now not be ready to overthrow him. However, the regular officer corps and influential citizens may be more willing to attempt to modify his behavior.
- It should be noted that without US intervention, 10 attempts to remove Qadhafi by members of his military have occurred since 1980 (*Los Angeles Times*, 27 April 1986).

The reception of the raid by the countries of the world was to some extent known before the raid and should be judged more as action taken than in words spoken. The European allies, recipients of most terrorist acts carried out against the West, are always fearful of additional acts. Since the raid, there has been unprecedented security and the sharing of antiterrorist information. An economic summit took place in which terrorist activities were discussed and meaningful ground rules were set up by the Western allies.

The raid on Libya appears to have been extremely successful as a psychological operation; yet it is doubtful that it was originally planned as a psychological operation. While any operation can have a psychological component, often that component is not a major consideration in planning and its effects are underestimated. If the raid had been originally planned as a psychological operation, the attack on Qadhafi's personal dwelling would have had to remain on the target list. But the tactical air strike would not need to be as accurate in a PSYOP mission. For example, the use of sonic booms and attacks on less-defended targets would most likely have produced the same results. However, in the final analysis, if senior planners considered the Libyan raid as a psychological operation, it achieved all its goals.

The Role of Military Psychological Operations in Support of National Antidrug Policy

Maj James V. Keifer, USAF, Retired

> *War does not belong in the realm of arts and sciences; rather it is part of man's social existence. . . . Politics, moreover, is the womb in which war develops.*
>
> —Carl von Clausewitz

The United States is engaged in a war against a drug culture that threatens to destroy the nation's social-political-economic fabric. The war is being fought on both domestic and foreign fronts against a culture of users and providers of illegal drugs. The national drug control strategy targets both supply and demand. National policy statements by Congress in the Anti-Drug Abuse Act of 1988 and by the president's 1990 strategy statement imply this to be a war of annihilation, but limited two- and 10-year goals make deterrence more realistically the objective of this strategy—deterring illegal drug use and trafficking. The US military, as an instrument of national policy, has been brought actively—although with definite limits—into this engagement. To fight the war, the US military establishment has decided—or has had decided for it—to avoid using one of its most potent weapons: PSYOP. Ironically, the US government believes it is important to bring other psychological instruments on both domestic and foreign fronts. If this country expects to achieve victory over illegal drugs, biases against the use of US military PSYOP should be recognized and put aside.

Is the nation truly at risk? What is the national antidrug policy, and what are its priorities, goals, tools, and strategies? What is the role of the US military in pursuing national antidrug policy? How can US military PSYOP be employed to

295

support the war against illegal drugs? To answer these questions, this essay examines how our national leaders define the threat. It also reviews recent drug policy, strategy, goals, and objectives, concentrating on those roles in which the military and PSYOP are most likely to become engaged.

The Threat Defined

National Security Decision Directive (NSDD) 221 of 1986 defines international drug trafficking as a national security threat to the United States. With the signing of this directive, the president placed illegal international drug trafficking on the official agenda of the National Security Council (NSC). This action established a requirement for the secretary of state to coordinate, through the chiefs of missions, all drug-related assistance, information, and activities within foreign countries. The State Department also served as the link between Central Intelligence and insurgency. The Director of Central Intelligence (DCI) was assigned responsibility to collect, analyze, and disseminate intelligence on the trade of illegal drugs. The secretary of defense was tasked to manage military drug-related activities, including plans, operations, intelligence support, drug interdiction, training of foreign military counterparts, and other related support.[1]

Even with the threat established, national policy and objectives were ambiguous at best. Activities reflected need for a national mechanism to coordinate interagency efforts and to provide expert, professional guidance. The political community saw the need for a clear-cut policy to attack the drug problem and put substance behind rhetoric. Things began to move—in the terms of Clausewitz, "If war is part of policy, policy will determine its character. As policy becomes more ambitious and vigorous, so will war."[2]

The President of the United States and the US Congress declared war on drugs with the passing of the Anti-Drug Abuse Act of 1988.[3] This legislation established the National Drug Control Program (NDCP) and created the Office of National Drug Control Policy (ONDCP) within the Executive Office of the President, effective January 1989.[4] William J. Bennett was named director of ONDCP to orchestrate the activities of participating agencies that were contributing to the combined effort. His position was designed to provide guidance, consistency, and continuity, and to resolve jurisdictional disputes.

Bennett's assignment to head ONDCP, leading a serious combined interagency effort, and to implement a comprehensive program to fight a burgeoning social-economic-medical problem, signaled an increased resolve in antidrug policy. Legislation directed the ONDCP director to set policies, objectives, and priorities for the NDCP and to provide the annual national drug control strategy for the president's presentation to Congress by 1 February each year. Congress required the strategy to include both long-range goals and short-term (achievable within two years) measurable objectives. Further requirements were to (1) define the balance between expenditures for supply reduction and demand reduction and (2) submit a review of state and local government efforts to support the program.[5]

National Drug Policy

Responsible directly to the president for recommending changes in the program's organization, management, and budget, the director of ONDCP serves as the executive branch representative on drug issues before Congress. He or she coordinates and monitors the implementation, execution, and fulfillment of the program's policies, objectives, and priorities by designated agencies. When agencies' policies fail to meet the responsibilities required by the national drug control strategy, the director notifies the agency head and directs that measures be taken to comply. The director also serves as chief consultant to state and local governments on national drug control matters.[6]

The Anti-Drug Abuse Act of 1988 provided the director with three principal subordinates: a deputy director for demand reduction, a deputy director for supply reduction, and an associate director to head a Bureau of State and Local Affairs. The president was responsible for nominating appointees to these positions. When confirmed by the Senate, they were prohibited from simultaneously holding other federal government positions.[7] The act recognized "the magnitude of the illicit drug problem and the threat it poses [to the] national security of the United States."[8] Disavowing proposals to decriminalize drugs, the act went on to say that "legalization of illegal drugs . . . is an unconscionable surrender in a war in which, for the future of our country and the lives of our children, there can be no substitute for total victory."[9] It then proclaimed, "It is the declared policy of the United States Government to create a Drug-Free America by 1995."[10]

The primary executive branch departments and agencies responsible for illegal drug supply reduction activities include the departments of Justice, Treasury, State, Transportation, Agriculture, Health and Human Services, Interior, and Defense; the Central Intelligence Agency, the Agency for International Development, the United States Information Agency (USIA), and the United States Postal Service. Those designated as primaries for demand reduction activities include the departments of Health and Human Services, Education, Housing and Urban Development, Labor, and Veterans Affairs. Additionally, the judicial system and various state and local agencies support the NDCP activities.

As the lead federal agency for the war on illegal drug trade, ONDCP was tasked to outline the national strategy, establish tangible short- and long-term goals, and coordinate and monitor those departments, agencies, and bureaus designated to take the battle to the enemy. The office published *National Drug Strategies* in September 1989 and January 1990. These documents served as the president's reports to Congress on the status of activities and intentions of the National Drug Control Program.[11]

Prior to establishing the ONDCP, the Executive Office of the President had promulgated the *National Drug Strategy and Implementation Plan of 1988*. This document identified six presidential goals: drug-free workplaces, drug-free schools, expanded treatment, improved international cooperation, strengthened drug law enforcement, and increased public awareness and prevention. Strategy focused on reducing the supply of illegal drugs, both grown in and imported into this country, and on reducing the demand for illegal drugs in the United States. The plan characterized the drug problem as multidimensional, involving medical, legal, security, economic, social, and educational aspects. The threat was defined in terms of drug type, production source, production increase, international cartel control, use of high technology, increased violence, and high-profit/low-risk crime.[12] President George H. Bush, in a September 1989 televised address to a nationwide audience, introduced the first strategy produced by the Office for National Drug Control Policy. His presentation outlined domestic and foreign objectives of a multifaceted attack on illegal drugs. On the domestic front, the president said the battle would be fought through programs of deglamorization, education, rehabilitation, and punishment. Foreign measures centered around programs of eradication, interdiction, extradition, and certification.

In his introduction to the 1989 *National Drug Control Strategy*, the director acknowledged that the national policy as articulated in the Anti-Drug Abuse Act of 1988—"it is the declared policy of the United States Government to create a Drug-Free America by 1995"[13]—is admirable but unrealistic.[14] Bennett proposed in its place, "the highest priority of our drug policy must be a stubborn determination further to reduce the overall level of drug use nationwide—experimental first use, 'casual' use, regular use, and addiction alike."[15]

This report from ONDCP concluded that the situation concerning illegal drugs was deteriorating rather than improving. Crime, health, and economic factors pointed to a growing problem. (Evidence at the time—figures showing declining casual use—suggested that the problems were primarily caused by addiction.) On both the foreign and domestic fronts, illegal drugs were cited as being imminently available; and the problems were readily evident. Commenting on the need to put realism into the national antidrug abuse policy statement, Bennett charged that a de facto policy vacuum existed. He said it was time to accept the fact that the war on drugs, to be effective, would be protracted in both time and reach. Serious, coordinated efforts and significant resources were needed to get a grip on the demand for and supply of illegal drugs.[16] Saying it would lead to a national disaster, Bennett articulated his adamant opposition to proposals for legalizing drugs.[17]

While building on the strategy from the Executive Office of the President, the 1989 *National Drug Control Strategy* was more than mere restatement of the 1988 plan. Unrealistic goals were rewritten and valid strategies were reorganized. The 1989 document outlined the national priorities as: (1) the criminal justice system, (2) treatment programs, (3) education, community, and business actions, (4) international initiatives, (5) interdiction activities, (6) research, and (7) intelligence. As mandated by the Anti-Drug Abuse Act of 1988, two- and 10-year objectives were listed. A statement on the process of implementing the national priorities was submitted with funding requirements. State- and local-level legislation initiatives were proposed to add consistency to activities below the federal level. The report articulated the importance of comprehensive information management to the program.[18]

The strategy recognized the contributions of a comprehensive information management program to the success of the antidrug campaign. In Bennett's introduction of the strategy, he expressed that coherent and coordinated policy-making depended upon the timely sharing of information. "Our national policy must be to maximize the sharing and use of relevant information among appropriate government organizations and to minimize impediments to its operational use."[19] He focused on improving automated data-processing systems and command, control, and communications (C^3) networks. In this context, information management referred to the intelligence function and communications systems rather than programs to provide people with educational material or persuasive information.

While the 1989 *National Drug Control Strategy* outlined initiatives that targeted supply-side activities, no relevant clear-cut policy statement on them was articulated. To remedy this omission, Bennett stated in the 1990 report that "the policy of the United States is to disrupt, to dismantle, and ultimately to destroy the illegal market for drugs."[20]

The 1990 *National Drug Control Strategy* was essentially a continuation of the previous year's strategy. Its major goals included (1) restoring order and security to American neighborhoods, (2) dismantling drug-trafficking organizations, (3) helping drug users break their habits, and (4) preventing those who have never used illegal drugs from starting. The strategy again called for cohesive, multifront attacks on both the supply and demand sides of the drug problem. In addition, the program's purpose was (1) to reduce to the maximum degree possible illegal drug supply and availability, (2) to provide treatment for users of illegal drugs and to actively discourage people from becoming involved with illegal drugs, and (3) to provide robust enforcement of drug laws by holding traffickers, sellers, the money-laundering structure, buyers, and users accountable.[21] National priorities for the strategy were directed as follows:

1. strengthening the criminal justice system,
2. expanding and improving drug-treatment programs,
3. developing programs to keep illegal drugs out of education systems, communities, and the workplace,
4. pursuing international initiatives,
5. enhancing interdiction efforts,
6. building on the research agenda, and
7. coordinating the intelligence agenda.[22]

Both supply and demand aspects were targeted for attack by the criminal justice system through actions to deter the use of illegal drugs, to disrupt the trafficking network, and to arrest, prosecute, and punish drug criminals. The 1990 strategy reinforced the 1989 strategy, which called for dismantling trafficking systems by apprehending and prosecuting organization leaders and accomplices, and by taking away their illegally acquired wealth. The goal of additionally holding buyers, sellers, and users accountable was to make illegal drugs less desirable, more expensive, and harder to get.[23]

The Office of National Drug Control Policy recognized that the US could not expect to unilaterally win the war against illegal drugs. The 1989 international strategy was designed to disrupt and dismantle the multinational criminal organizations that support the production, processing, transportation, and distribution of drugs to the United States and to other nations. Building on that strategy, the 1990 programs were aimed not only at drug-producing, transit, and consumer countries, but also toward nations still experiencing few or no problems with illegal drugs by helping to strengthen resolve and resistance to use and trafficking inside their territories. The 1990 report articulated the need for international cooperation if objectives were to be met.

Three near-term goals the national strategy sought to attain in the international arena were as follows:

1. providing support to help strengthen the political will and institutional capabilities of three Andean Ridge countries (Colombia, Peru, and Bolivia) by enabling them to fight illegal drug-producing and drug-trafficking organizations in their respective countries,

2. helping law enforcement and military structures of the three countries to combat cocaine trade more effectively, and

3. incapacitating trafficking organizations that operate in the three countries. To ensure understanding of and to develop support for the measures, the 1990 strategy tasked that "U.S. information and public awareness programs will explain and support the attainment of the three goals outlined above."[24]

The 1990 strategy added a new dimension—the implementation of international public information initiatives to directly support other international policies and programs. "An active public information campaign will provide vital information to foreign publics, leaders, and government officials to build support for United States and host country actions to combat drug production, trafficking, and consumption."[25] Receiving policy guidance from the Department of State (DOS), the USIA (with support from other federal agencies) was assigned responsibility for coordinating and leading the international information effort. These initiatives focused on informing foreign audiences about the problems caused by illegal drugs with respect to their national security, economic welfare, and environment. International information programs were to educate foreign audiences about the consequences of using illegal drugs. Another initiative of the international information program was to describe the US domestic drug problem and the progress experienced in fighting it.[26]

The 1990 *National Drug Control Strategy* included a commitment to long-term research in the fields of treatment, education and prevention, criminal justice, and drug use.

Research provided a base of knowledge for fighting the drug problem and a foundation for future strategies. Research projects included these:

1. developing enhanced databases on drug production and consumption, and on the economic impact of drugs, including cost/benefit analyses of programs designed to reduce drug use;

2. developing regional and state drug-related data to improve the analysis and accuracy of assessing drug control effectiveness;

3. expanding domestic programs to collect information on illegal drugs and their use;

4. increasing scientific and technological research projects focused on strengthening drug enforcement and interdiction capabilities; and

5. increasing research into drug treatment to emphasize complications caused by addiction, acquired immunodeficiency syndrome (AIDS), and pregnancy.[27]

The intelligence agenda, designed to understand and fight illegal drugs better, included acquisition, analysis, and management of information. The National Drug Intelligence Center was tasked to coordinate, consolidate, and distribute pertinent information to appropriate agencies. Responsibility for coordinating and overseeing drug-related intelligence-gathering activities of the national foreign intelligence community was tasked to the DCI. The strategy required all members of the nation's foreign intelligence community to share data relevant to drug trafficking and drug-related money laundering. This included requiring DOD's intelligence components to actively support the collection of foreign drug-related intelligence requirements outlined in the *National Drug Control Strategy*.[28]

To effectively manage the National Drug Control Program, ONDCP established coordinating mechanisms, developed report and study requirements, and set up systemic measures to oversee priority objectives. The Supply Reduction Working Group, the Demand Reduction Working Group, the Research and Development (R&D) Committee, and the Drug-Related Financial Crimes Policy Group were created to coordinate interagency activities. Drug control program agencies were created within federal departments, agencies, and bureaus to manage the organizations' internal activities and serve as single points of contact for responding to drug issues. Lead agencies were designated to direct, coordinate, and provide expertise to specific activities. Management studies included the Executive Reorganization Study, the Department of Justice Reorganization Report, the R&D Facilities Plan, and reports on communications and automated data-processing initiatives. The Office for Treatment Improvement was created within the Department of Health and Human Services to lead the federal effort to improve the nation's drug-treatment activities. Information clearinghouses were set up in departments, agencies, and bureaus to access data readily. The National Drug Intelligence Center was established to oversee information management. Statutes were implemented at state and local levels to deny convicted drug offenders access to specific federal benefit programs. A programming and budgeting system was created within ONDCP. Model legislation on drug-free workplaces was provided to state and local governments.[29]

Information management initiatives—developing command, control, communications, and intelligence networks—were built on the previous year's strategy. The only notable difference was in the change of the terms from *communications* to *telecommunications*. The strategy again focused on upgrading hardware, integrating systems, and providing means for secure communications. The DOD, as executive agent for implementing communications systems, created the Counter-Drug Telecommunications Integration Office within the Defense Communications Agency to serve as the focal point for all telecommunications activities related to the enforcement of drug policy.[30]

An Assessment

Both the president (NSDD 211, 1986) and Congress (Anti-Drug Abuse Act, 1988) unambiguously articulated official positions that illegal drugs are a threat to national security. They agreed that the political objective, elimination of demand for and supply of illegal drugs, deserved an extensive effort and merited significant expenditure of resources. The problem was identified in global terms with an objective to co-opt worldwide support for counternarcotics initiatives. War was declared—not the conventional sense of a military shooting war (although blood most certainly was to be shed), but one of police actions, economic sanctions, information activities, social programs, and other nonmilitary means.

The enemy in this war is the drug culture—drug cartels and drug users. The drug cartels cannot be likened to the contemporary economic entities; a cartel is a flexible system of organizations involved in growing, producing, transporting, distributing, selling, and money-laundering processes. High profitability has provided them an immense resource base. The cartels are motivated by profit rather than ideology. Users of illegal drugs cut across the full spectrum of society, from the homeless to professionals to influential leaders. The group includes both casual and addicted drug abusers. Their motivations for using illegal drugs are equally diverse.

Just as the war against illegal drugs is not fought against a conventional enemy, it should not be considered an ideological conflict. It is unlike most other political wars including the War on Poverty waged by the Great Society during the 1960s. Because the drug problem transcends many cultural boundaries, care must be taken to consider relevant issues such as social, health, safety, and economic factors, and to avoid morality issues. In seeking international cooperation, the perception of selling American values to other countries must be avoided. Actions and propaganda should be controlled, so as to preclude being tagged "Yankee imperialism" or "American hegemony."

The closest parallel we have to this war is the battle against alcohol that took place during prohibition—a war that was lost but that certainly compromised national objectives. Proponents of legalization can be expected to argue the logic of prohibition repeal in their efforts to sway the public to support such measures. The reasons for prohibition's failure must be understood if history is not destined to repeat itself. Care must also be taken in arguing against legalization to avoid the pitfalls of incomplete truths and invalid logic that cannot stand the tests of intense scrutiny or time.

The Anti-Drug Abuse Act of 1988, in declaring that national policy was to make a drug-free America by 1995, reflected election-year rhetoric. Remarkably, the political process that created this legislation didn't prevent the promulgation of a generally comprehensive[31] and reasonably good product—but it did serve to dilute some of the potential it carried. While putting together a product calculated to appeal to the American public, members of Congress ensured that their power would not be diminished.[32] Appropriately, the legislation directed attacks against both supply and demand of illegal drugs.[33] It also stressed both counterforce and countervalue strategies. With respect to one strategy that has been suggested by some— the legalization of certain drug offenses—Congress refused to compromise.[34] On the positive side, the single-point manager established a mechanism for coordinating efforts and resolving disputes. The requirement that leaders of the ONDCP not hold any other public office kept the president from appointing part-time people.[35] On the negative side, the political nature of the legislation tied the effectiveness of the program to (1) its position on the president's agenda and (2) its proximity to election day.

Supply-side measures receive the attention of Congress, but the trafficking of illegal narcotics isn't their exclusive target. Prevention and treatment programs for this legislation demand side attention. Certification requirements focus on governments of countries in which drugs are produced and through which they transit. Measures to break up money-laundering schemes put pressure on a shadowy support structure that helps the cartels to reap the huge profits of their trade. Opposition to decriminalizing illegal drugs, while supportive of the national policy stated by Congress, will eventually generate the requirement to produce facts from research to support this stance.

The most notable shortcoming of the legislation is the comparatively little formal authority given the director's position, particularly in the light of its extensive responsibilities. The director has few immediately assigned resources and only limited fiscal control. While Congress is expected to keep its control over the purse, authorizing the director to manage the distribution of confiscated assets/resources would provide the position with a significant carrot for resolving turf battles.

Most of the combatants in the drug war are assigned to the numerous supporting departments, agencies, and bureaus. With no formal control over them, the director must resort to convincing, persuading, or otherwise getting these supporting elements to do the work according to plans. Resolution of problems and conflicts tends to be a matter of Capitol Hill politics,[36] a personality-dependent requirement ideally suited to a person with good diplomatic skills. Disagreements of consequence must be resolved by the president or significant others in the Washington community. Further, few would argue that the director's position holds stature equal to such leaders as the secretary of state, secretary of defense, attorney general, or national security advisor. Not a formal member of the cabinet or the NSC, the director sits in only when the topic of illegal drugs is on the agenda.

Articulating the national policy for demand-side measures, Bennett's September 1989 *National Drug Control Strategy* changed Congress's declaration from unrealistic (drug-free America by 1995) to inconsequential (reduce overall level of drug use). Only one of the strategy's two- and 10-year goals[37] serves to measure the ultimate objective of the national policy statement for supply reduction to destroy the illegal market for drugs. The objectives and strategies formulated by the ONDCP represent a significant step forward from the Just-Say-No! domestic policy and interdiction instruments of the Reagan administration. Recognizing the complexity of the problem, the strategy addresses supply and demand, domestic and foreign arenas, law enforcement and prevention and treatment programs, counterforce and countervalue measures, and casual use and addiction. Quantified objectives are provided to measure some effects of the program. The two- and 10-year goals appear to be realistically achievable, even possibly unambitious. (Figures show a steadily diminishing demand for illegal drugs.) Bennett's uncompromising position on decriminalization is based on emotional grounds. Research needs to be directed to prove the validity of this position; that the effects of the drugs themselves, rather than their criminal status, have accelerated and amplified social, economic, health, and safety decay.

The strategy outlined in the reports by ONDCP has many strong points. It states that one strategy alone will not win the war. Efforts to reduce the demand cannot be expected to work without measures against supply production and vice versa. The high profitability of drug trafficking will lead to aggressive marketing efforts by the cartels to counteract gains of demand-side programs. Interdiction efforts may thwart transportation networks and make smugglers think twice before trafficking the products, but cartels will fight these measures by cultivating new markets, developing new transportation strategies and transit routes for their products, and increasing rewards to their couriers. By the same token, supply-side measures alone can be unsuccessful. As long as drugs are available to tempt people, a portion of the population will use them. Treatment programs are extremely costly and none has escaped problems with recidivism. Incarceration is likewise costly and has hardly proved effective when not supported by other measures.

Research and development (R&D), essentially an investment process, is the primary way this nation will come up with different and improved ways to fight illegal drug use. It serves as the tool for innovating new ideas and technologies. The commitment to R&D acknowledges the long-term nature of the war against drugs. One of the foremost obstacles the ONDCP faces is a dearth of accurate, up-to-date information concerning illegal drug use. Previous surveys were conducted too infrequently and were of limited or irrelevant scope. They lacked sufficient scientific rigor to be reliable measures. This was one of the primary factors for understated program goals of the initial strategy published in September 1989. Funding for databases will enable them to be expanded and updated more frequently, increasing both the reliability and utility of the information gathered. Some R&D projects are being designed to improve the effectiveness of prevention and treatment programs by helping to reduce the demand side of the equation. Other programs are being advanced to improve technology for detecting drugs and for systems to compile, communicate, and coordinate information to help supply-reduction efforts.

Recognition of the need for international consensus is an important element of the program. Not only are bilateral agreements needed with drug production and transit countries to help them fight their internal enemies, but worldwide cooperation is also an instrumental part of finding a solution to this global crisis. Crop eradication projects will still take place, but the program has been deemphasized because of some counterproductive effects. (As drug-producing plants constitute the primary cash crop in many Andean Ridge countries, many farmers side with revolutionary factions, creating additional problems for their governments.)

Another key element to the strategy is the importance given to information, education, and publicity programs (essentially, programs of advocacy under politically palatable titles) on both domestic and foreign fronts. The success of information strategies relies on communicating in understandable terms the consequences of illegal drug activity. Some initiatives are designed to persuade nonusers not to start, users to seek treatment, and traffickers to cease their activities. Because international understanding and backing are imperative to the ultimate success of this program, other initiatives seek support from the world community for our national objectives.

The formation of coordinating committees is another program strength that promotes the sharing of ideas, resources, and information.[38] These committees establish mechanisms to prevent one agency from unwittingly doing something that conflicts with other efforts. This initiative improves cost-effectiveness and prevents potential fratricide. A notably missing committee is one to coordinate publicity, information, education, diplomacy, and persuasion activities of federal, state, and local agencies.

The strategy is not without its share of weaknesses. Because it is a long-range program, it will test the national patience of people who are accustomed to instant gratification. This constituency in turn will challenge the resolve of politicians whose primary concern is their own reelection. These same policy architects, if they don't perceive the threats—the threat drugs pose, and threats to reelection—as imminent, will most likely prove unwilling to commit sufficient assets over the long run to achieve success.

While expenditures have increased dramatically, the total program budget compared to low-end estimates of profits made by drug traffickers gives the impression that the administration is still trying to win votes rather than the drug war. This is particularly true because funding is dedicated to supply-reduction measures. Less than one-fifth of antidrug funding goes to prevention and treatment programs. Funding for prevention and treatment, as well as alternative programs, gets the short end of the stick when compared to the amount given to the interception of smuggled goods.

The strategy for providing support to state- and local-level government programs relies on the provision of matching funds by the lower levels to qualify for the federal money. This reliance depends not only on the availability of funds for these purposes but also on political popularity. To attract popular support to raise the taxes needed to fund these programs, results will have to be visible with high payoffs.

The inclusion of measurable objectives is laudable, but the selected objectives—with the noted exception of drug availability reduction—tend to measure only the demand side of the program. While this helps to dissuade "body-count" tactics, these objectives fail to measure the effectiveness of component strategies and programs.

The national policy of interrupting cartels as a departure point to their eventual dismantling and destruction implies a short-term goal of harassment. It serves to challenge cartels to produce new ways to circumvent measures. The absolute nature of the national policy—ultimately to destroy the illegal market for drugs—represents the objective of annihilation. If this is not wishful thinking but the true goal, then half-hearted measures will fail in the long run.

Bringing the military into the counternarcotics strategy was an expedient decision for the political leadership, but one the military leadership looked upon with skepticism. The political side saw it as a way of putting defense dollars to a tangible use. Political reasoning generally followed this theme: while ships and planes are out on training missions, they can be on the lookout for drug smugglers. The military establishment saw itself being dragged into a war with no measurable objectives; a war that could not be won. Through appropriations bills, Congress provided funds to fight the drug problem, thus easing military apprehensions about taking the costs out of the operating budget. Defining the Defense Department's fiscal authority for 1990, Congress dedicated $450 million to counterdrug activities.[39] This funding included $28 million for R&D projects and $40 million for catchall expenses defined as additional support to the Office for Drug Control Policy.[40, 41] The bulk of the spending authority, $182 million,[42] was to support interdiction in its lead role of detecting and monitoring aerial and maritime smuggling activities. While the added fiscal authority generated interest, the lack of significant measures of effectiveness make military counternarcotics activities a readily available political and media target. Coupled with operating restrictions, in spite of the loosening of the *posse comitatus* statute, DOD resources offer more potential than support. Its role as lead agency in the C^3 arena should easily show more cost-effectiveness than interdiction responsibilities when networks are completed and figures come in.

If national objectives in the war against illegal drugs are met, the achievement will ultimately be a psychological one. Activities must be carried out on both domestic and foreign fronts. They must target both supply and demand cultures, support counterforce and countervalue measures, and address prevention and treatment. National resolve must be communicated in clear and concise terms. Persuasion will require both action and propaganda.[43]

For political reasons and other expediencies, these messages will have to be carried to their audiences by many different agencies. The 1990 strategy announced campaigns to provide information to foreign audiences. In the Anti-Drug Abuse Act of 1988, Congress established the requirement for the Office of National Drug Control Policy to activate a program to publicize the penalties for violating the legislation.[44] Agency turf battles during World War II and the American cultural distrust of the psychological instrument led to prohibitions against the USIA's communicating directly to a domestic target audience. Similarly, the *posse comitatus* statute and other restrictive measures prevent US military resources from conducting activities intended to influence US citizens. Domestic programs, therefore, must be carried out by other agencies.

In the foreign arena, the psychological dimension must be exploited by all agencies involved in activities. Actions must be reinforced by publicity campaigns to ensure that maximum benefits are reaped. Successes must be underscored and punctuated, letting others know that alternatives exist and that consequences are in store for unacceptable behavior.

Co-opting the support of nongovernment special interest groups is another potential role for the psychological instrument. Informing the public of the harm illegal drugs do to the environment, the economy, the health, and ultimately the strength of a nation could serve to coalesce and polarize "watchdog" groups, directing outrage and magnifying public opinion against cartels and their support structures. For instance, environmental groups such as Greenpeace might find that the rape of the Andean Ridge ecosystem is a target for one of their campaigns. The focus of attention by television programs such as "60 Minutes" on chemical companies that exercise irresponsible management of precursor materials might cause them to rethink their motives for profit and to tighten control mechanisms.

The top echelon of the cartel structure is not invulnerable to psychological inducements. For all their wealth, they are virtual prisoners, unable to attain status or to travel any appreciable distance from their fortified homesteads. Psychological initiatives can highlight and amplify the social isolation these people face. Rivalries between cartels can be exploited, as can rivalries between ideology-motivated revolutionary movements and the greed-motivated cartels.

The across-the-board rejection of any form of decriminalization places an even greater demand on the psychological dimension. Diplomacy has been described as a process of political compromise between nations.[45] Because the drug culture has no national identity and because compromise has been determined unacceptable, diplomacy would seem to be limited to securing international cooperation from nations predisposed to strong antidrug policies. The goal is to convince people not to use illegal drugs, to convince cartels that it isn't in their *profit-motivated* interest to traffic illegal drugs, and to

312

convince the world community to act together to stop the trafficking of illegal drugs. Two overriding principles should guide the psychological effort: (1) truth is the best propaganda; and (2) words must follow action.[46]

To support the first principle, the government needs to intensify its R&D studies to come up with irrefutable reasons not to decriminalize drugs. These studies should compare health and safety, sociological, and economic factors upon which illegal drugs are claimed to have an impact, comparing states and nations that have and have not decriminalized drugs. This type of research should then establish a mechanism to control the impact of the criminal status of drugs. Emotional appeals will only have temporary effects on the behavior of their target audiences. A comprehensive interagency campaign using facts should be developed by a coordinating committee. Likewise, morality inferences concerning drugs must be carefully tailored to the morals of target cultures.

While it has been written that words are mightier than the sword, they must be based on substance. The principle that propaganda must follow action recognizes that the propagandist shouldn't make claims he is not willing to back up or promises he can't keep. Credibility is ideally established through performance of actions that serve as examples of what may be yet to come. This performance establishes both capability and resolve. If the government seeks to alienate cartels from their surrounding populaces by promising to reward and protect informants, it must be prepared to back up its words or suffer from challenges to its resolve.

Alternatives must be either available or provided. For instance, coca-growing campesinos cannot be expected to voluntarily stop growing a profitable cash crop without the assurance they will receive a reasonable price for harvesting something not considered illegal. Likewise, others throughout the cartel structure must see some viable way to generate an income if they are to be induced to get out of the drug-trafficking business.

Where then does the US military PSYOP community fit into the equation? First and foremost by ensuring that the actions of DOD (maritime and aerial interdiction and detection; monitoring and developing C^3 networks) are favorably

313

communicated to the world community. This supports the premise of following actions with words—making would-be smugglers aware of the diminished likelihood of their success and profit, and broadcasting the nation's resolve to stop drug trafficking. Another role is in working with foreign counterparts to help them master the use of psychological instruments against illegal drugs. For countries whose military forces are used to combat drug traffickers, US military PSYOP forces can conduct seminars on the value of psychological operations, train their contemporaries, and participate in bilateral exercises to hone skills. For those countries whose enforcement agencies are civilian, US military forces can support USIA programs of assistance.

Direct measures to communicate with foreign civilian populaces, while not out of the scope of possibility, are more appropriately the responsibility of the USIA. That doesn't preclude a military PSYOP role but suggests coordination of ideas, information, and efforts. This role can be particularly important because the lead agency suffers from a significant funding lack, given its responsibility to execute the national strategy.

If the US government is to destroy the market for illegal drugs successfully, the psychological dimension will be a significant factor. Military PSYOP resources can contribute to achievement of the national objectives, but only if they are allowed to be used. To that end, institutional biases will have to be broken. Ironically, such a breaking of biases is a psychological phenomenon. This process will include overcoming Western cultural fears of manipulation, resolving interagency turf battles, motivating the military establishment to consider nonlethal weapons, and educating the members of the media. These audiences must be made to understand that psychological activities are amoral,[47] that they support national policy, and that advances in technology have made far-reaching communications faster and stronger.

Dyer states the need for a national college to educate leaders on the nature of the instrument and how it helps them to implement policy.[48] The public, he says, "must be educated on what these activities are and how they contribute to functioning of the government."[49] Within the military community, education on the psychological dimension needs to be expanded. In

314

addition to courses of instruction at the Air Force Special Operations School at Hurlburt Field, Florida, and the US Army John F. Kennedy Special Warfare Center and School at Fort Bragg, North Carolina, blocks of instruction should be included in service academies and in professional military education courses at all levels. The Defense Information School at Fort Benjamin Harrison, Indiana, should include psychological operations orientation in its military public affairs training curricula. Select members of the media should be invited to attend military orientation courses.

Bennett resigned from his position as director, commenting that he had accomplished what he had set out to do. Acknowledging that the war was not yet won, he offered that indicators were showing great promise.[50] His abrupt departure provided focus on a national program whose status was being subordinated by events in the Middle East. The war is still far from over, and it will not be over until the national policy is achieved or rescinded. While national policy is likely to undergo redefinition, particularly as accurate and up-to-date information enhances the nation's databases, rescission is unlikely. Changing peoples' attitudes, opinions, and behavior will be a long-term psychological process—a process to which the military PSYOP community has much to offer. The sooner political and military leaders recognize this, the sooner the problem will be resolved.

Notes

1. Betac Corporation, *Department of Defense Counterdrug Baseline Report* (21 July 1989), 2-10 through 2-11.

2. Carl von Clausewitz, *On War*, ed. and trans. Michael Howard and Peter Paret (Princeton, N.J.: Princeton University Press, 1976), 606.

3. US Congress, *Anti-Drug Abuse Act of 1988*, Public Law 100-690, 100th Cong., 18 November 1988. Sec. 4801 states that problems caused by illegal drugs pose a threat to the national security of the United States.

4. Ibid., Title I, Subtitle A, sec. 1002, 102 Stat. 4181.

5. Ibid.

6. Ibid., Title I, Subtitle A, sec. 1003, 102 Stat. 4182–83.

7. Ibid., Title I, Subtitle A, sec. 1002, 102 Stat. 4181.

8. Ibid., Title IV, Subtitle I, sec. 4801, 102 Stat. 4294.

9. Ibid., Title V, Subtitle A, sec. 5011, 102 Stat. 4296.

10. Ibid., Title V, Subtitle F, sec. 5251, 102 Stat. 4310.

11. Ibid., Title I, Subtitle A, sec. 1003, 102 Stat. 4182.

12. Ibid., Title I, Subtitle A, sec. 1005, 102 Stat. 4185–86.

13. Public Law 100-690, Title V, Subtitle F, sec. 5251, 102 Stat. 4310.

14. Office of National Drug Control Policy, *National Drug Control Strategy* (Washington, D.C.: Government Printing Office, 1989), 9.

15. Ibid., 8.

16. Ibid., 1–4.

17. Ibid., 9.

18. Ibid., 43–44, 84–90, and 131–36.

19. Ibid., 93–97.

20. Office of National Drug Control Policy, *National Drug Control Strategy* (Washington, D.C.: Government Printing Office, 1990), 1 (hereafter cited as ONDCP 1990 Strategy).

21. Ibid., 1–9.

22. Ibid., 12–85.

23. Ibid., 12–27.

24. Ibid., 51.

25. Ibid., 59.

26. Ibid.

27. Ibid., 74–81.

28. Ibid., 82–85.

29. Ibid., 107–16.

30. Ibid., 123–28.

31. One hundred eighty-seven pages detailing everything from American Indian cultural exceptions to penalties for the use of firearms in conjunction with a criminal drug offense.

32. Congressional oversight committees provided the director power to influence but not control their agencies' activities.

33. Public Law 100-690, Title V, Subtitle F, sec. 5251, 102 Stat. 4309.

34. Ibid., Title V, Subtitle A, sec. 5011, 102 Stat. 4296; and Title VI, Subtitle F, sec. 6201, 102 Stat. 4359.

35. Probably most noteworthy, the vice president, the office that ran this program under the Reagan administration, is excluded.

36. Capitol Hill gamesmanship is well covered by Morton Halperin's *Bureaucratic Politics and Foreign Policy* (Washington, D.C.: Brookings Institution, 1974).

37. ONDCP 1990 *Strategy*, respective 10 and 50 percent reductions in drug availability, 120.

38. ONDCP 1990 *Strategy*, 107–16.

39. Defense Authorization Act for 1990–1991, 103 Stat. 1562, Public Law 101-89, 29 November 1989, Title XII; and Public Law 101-65, 21 November 1989, Title VII; Stat. 1128.

40. Ibid., sec. 1205.

41. Ibid., sec. 1202.

42. Ibid., sec. 1204.

43. Murray Dyer, *The Weapon on the Wall* (1959; reprint, New York: Arno Press, 1979), 35–37. Mr Dyer's fifth premise, words are rooted in action. Action demonstrates capability and resolve; propaganda communicates and intensifies the actions' effects.

44. Public Law 100-690, Title V, Subtitle A, sec. 5011, 102 Stat. 4296.

45. Martin Wright, *Power Politics,* ed. Hedley Ball and Carsten Holbraad (Leicaster, U.K.: Leicaster University Press, 1978), 27.

46. From premises outlined by Dyer in chap. 2.

47. Dyer claims they "cannot make the necessary decisions . . . only explain and interpret them," 213.

48. Ibid., 187–90.

49. Ibid., 199.

50. The price of cocaine had risen sharply and street supplies were down.

Political-Psychological Dimensions of Counterinsurgency

Gen Richard G. Stilwell, USA, Retired

This paper concentrates on the psychological aspects of conflict because the psychological dimension deserves discrete treatment as the least appreciated dimension—and the least employed—by the United States and most of its allies. Conflicts erupt among groups of men, nations, or coalitions of nations from collision of aims or objectives for which the antagonists are prepared to fight and die. The conflict ends, at least temporarily, when one side makes the decision that there is more to be gained—or less to be lost—by allowing the antagonist to prevail. The side that desists has simply lost the will to continue the conflict.

An adversary's will can be eroded and broken in a number of ways: by military force (the application, or threat of application, of violence); by economic strangulation; by loss of external support; by apathy; and by military defections. Normally, it takes a combination of these to bring about an adversary's calculus that the "jig is up." Sometimes, one master stroke will suffice. Hopefully, nations can resolve the issue without the application of military force. To win without fighting, in the words of Sun Tzu, is "the acme of generalship." Any political leader or military commander should accomplish the mission in the most expeditious fashion and at minimum human cost.

The term *psychological warfare* (psywar), for all its short-comings, succeeds in combining two meaningful subjects, both as old as man. Certainly, it is better than *psychological operations* which, as my mentor Paul Linebarger once said, leaves the issue entirely neutral. Psychological warfare seeks to achieve the objective where military force is unavailable or inappropriate, or where it can combine with the military to minimize expenditures while maximizing effects.

The Old Testament's "Book of Judges" provides one of the earliest and most effective examples of psychological warfare. Gideon—an imaginative fellow—and his troops were on the verge of being annihilated by a Midianite force of vastly superior strength. The Midianites were encamped in preparation for decisive assault on the morrow. But Gideon had an inspiration: Since the basic fighting unit of all armies in the thirteenth century B.C. was a 100-man formation, each with one trumpeter and one torch bearer, he reckoned he could create the impression of a 30,000-man force with 300 men properly equipped. Gideon selected 300 men and provided each with a lamp, a water pitcher, and a trumpet (we are not told where he found the extra trumpets!). At nightfall, he had the 300 light their lamps (which were placed inside the pitchers so as to hide their light) and deploy at designated intervals around the Midianite camp. On his signal, they all broke their pitchers, lighting up the perimeter, and blew madly on their trumpets. The Midianites were startled out of their sleep and their wits. They fought among themselves, then gave up and retreated. (The Hebrew chronicler gives credit for this to the Lord!)

Gideon's deception is not an isolated example, as any serious student of ancient military history can confirm. Psychological warfare seeks to persuade by nonviolent means. By its very nature, psychological warfare is open-ended; it defies an accurate all-encompassing definition. The best descriptor—although by no means all-inclusive—is propaganda, which unfortunately, has severe political handicaps on the American domestic scene. It ought not be in such disrepute; after all, it derives from the name of that department of the Vatican Curia which had the duty of propagating the faith. Propaganda is simply the *planned* use of any form of communication designed to affect the minds, emotions, and actions of a given group for a specific purpose. This is precisely the aim of what we see on the television screen at all-too-frequent intervals—a field of endeavor in which the United States is preeminent.

So, psychological warfare is really not all that esoteric or mysterious. Whether in support of the national security or for the profit of shareholders, the effectiveness of propaganda depends on adherence to four basic ground rules: a clear-cut aim or purpose, a well-defined target audience, a credible message, and a reliable means of communication or dissemination. Gideon met these criteria in spades. His aim was to demonstrate that he had been greatly reinforced; his target audience was the immediately opposing army; the established rules of counting made his message credible; and the light and noise ensured that the message was received. To be sure, it is rarely that easy; or as quickly mounted; or as clearly measurable.

The American View of Psychological Warfare

The psychological weapon has never been accepted as a permanent instrument of national security policy by the executive branch, the Congress, or the American people. Its use, under various names, has been supported as an essential expedient only in periods recognized as war. To be sure, in the early days of the cold war, there was considerable enthusiasm for institutionalizing a national structure and capability to exploit international communications as a psychological tool in furtherance of national policy.

The concept was never implemented, due in part to substantial differences within the executive branch about the value and propriety of this genre of activities. These differences precluded the enunciation of an integrating doctrine and the necessary coordination of departmental programs. In greater part, however, the failure to implement was due to the American character, which is at once idealistic and pragmatic, distrustful of political intrigue, and impatient for quick solutions. These traits are reinforced by the rather simplistic and innocent view of the world that most Americans have.

For these reasons, instances of US attitudinal readiness—and US capabilities—to engage in psychological warfare have been rare, coinciding with the two world wars. Those efforts were indeed national because the target audience conspicuously included the American public itself. But while psychological operations in the two world wars (particularly the British in World War I) provided important support, they were not crucial to the final outcome. Those conflicts were waged militarily, with political and psychological assistance.

Wars of national liberation reverse the weighting of factors. They are fought politically and psychologically, with the assistance of military capabilities. They integrate political, psychological, social, and military aspects to exploit national vulnerabilities, erode all national institutions, and support an eventual insurgent takeover. Their choice of weapons puts a conventional high-technology force at a disadvantage. The psychological challenge is explicit in the hallmarks of the insurgent force: its tight, inner political structure; the strong motivation of all ranks; the ability to find concealment within the civil population; and identification with one or more popular causes.

Past Experience in Counterinsurgency

The wars of national liberation that have erupted in the third world over the past four decades document the crucial role of the psychological weapon in determining the final outcome. Not all involved the United States, even indirectly; but all were replete with useful lessons for those perceptive enough to appreciate and assimilate them. In those instances where the United States was in some way engaged, its record in exploiting the psychological dimension is uneven. Two examples will serve to support this judgment.

The Philippine government campaign against the Huks in the early 1950s stands as a model of the imaginative employment of psychological warfare, strategically and tactically. The charismatic and insightful Secretary of National Defense, Ramon Magsaysay (who was later to become president), and a small cadre of exceptionally gifted US advisors, achieved spectacular success in arresting the momentum of the insurgency and then breaking its back with a wide-ranging series of programs which integrated military, political, and psychological actions. Most of the ideas and initiatives were conceived by the advisors, but—and this is important—the detailed planning, the message content, and the operational execution were all Filipino.

Magsaysay knew instinctively that the key to defeating the guerrillas was to deprive them of support within the rural population and to swing that support to the government. He also recognized that the soldier was the most visible symbol of government in the countryside and that the government would be judged by the actions of its soldiers.

As a first order of business, therefore, Magsaysay set about reindoctrinating the entire Philippine army. Its multiple roles were protector of the people, guerrilla fighter, and contributor to the morale and welfare of the civil population. He made this indoctrination meaningful by promoting only those who were demonstrably effective ambassadors of good will and were also effective combat soldiers. He provided rations for the troops that were adequate not only for their own consumption, but also for donations to villagers in dire need, and he made extensive use of medical and engineer personnel for civic actions of the most basic sort. Finally, Magsaysay established a "hotline" to his immediate office for complaints about troop malperformance and a staff to immediately investigate those complaints in the field, convincing one and all that he was serious about making the army a credible symbol of good and caring government.

Every effort was made to publicize, nationwide, the new order in the army and what was being accomplished thereby. To accomplish this, Magsaysay used press releases, unit newspapers for troop consumption (and for subsequent distribution to local civilians), movies of operations and civic actions (for showing in remote areas by civil affairs officers), and traveling road shows. Programs to win the support of civilians were complemented by others directed at the insurgents themselves.

One such program provided surrenderees (and even those who had been captured) the wherewithal to start life anew for self and family: substantial acreage in a resettlement area and a loan of building materials, tools, food, and seed. This program stimulated many surrenders, but its indirect effect was even greater. Much of the population had supported the guerrillas because their avowed motive was to gain "land for the landless." When it became apparent that the objective could be achieved without fighting, the moral obligation to support the guerrillas disappeared.

A second program involved liberal rewards for the death or capture of Huk leaders who were brought in or killed by their own comrades. The principal values of this program lay in widening the gulf between civilians and guerrillas, and in heightening hostilities within the guerrilla ranks. Accompanying these major programs were more specifically targeted initiatives: appeals from mothers to their guerrilla sons; rumors to destroy the credibility of those politicians who were deliberately obstructing the Magsaysay programs; widespread distribution of posters exposing Huk leaders as criminals wanted for documented murder, kidnapping, and rape; and exploitation of native superstitions.

The concomitant of steadily increasing popular support for the government—as symbolized by the army in the field—was better intelligence and more effective operations against the guerrillas. By 1953, the insurgency was no longer a menace to the national security of the Philippines.

What lessons can be drawn from the significant contribution of psychological warfare to the success of the Philippine counterinsurgency campaign? One is that Magsaysay had his priorities right. In most third world countries, the soldier in the field is the embodiment of the government. His conduct will determine the nature of the relationship between the population and the government in rural areas. The soldier must therefore be made the most useful symbol possible, demonstrating the moral justification for government. It is a nondelegable command responsibility, at all levels, to ensure this.

Another is that several factors combined to swing popular support to the government through actions of the Philippine military:

1. leadership and command emphasis (notably including careful indoctrination of the soldier);

2. a dedicated element in each headquarters to plan and supervise implementation;

3. psywar programs carefully tailored to the attitudes and needs of the groups at whom targeted; and

4. quality advisors with the professional competence, empathy, and passion for anonymity to gain and maintain access to the decision makers.

No one spent any time attempting to define boundaries of, or responsibilities for, conduct of psychological operations. It was everybody's business to support the overarching aims: develop confidence in the government and render the insurgency futile. This was achieved by the remarkable integration of concrete and useful actions, increasingly effective field combat operations, and support from information/psywar initiatives.

Vietnam was an entirely different story. There was an illuminating exchange between Col Harry Summers and a North Vietnamese officer in Hanoi during the prisoner of war exchange negotiations in 1973. Summers made the point that the American army had never been bested in battle to which the response was "Quite true, but that is totally irrelevant." The communists' political and psychological campaigns were decisive—on the international scene, within the American body politic, and in-country. The tragedy is that our failure to mobilize international and domestic support for the Nicaraguan Freedom Fighters mirrored our Vietnam experience in those key arenas.

A comprehensive and authoritative account of psychological operations in Vietnam has yet to be compiled. When accomplished, it will record some brilliant successes at the tactical level and some failures at the strategic level. The latter included an inability to cope with the impact of modern mass communications (history's first televised war); the lack of psychological/political actions to condition audiences and establish context for major military activities as, for example, the invasion of Cambodia; inability to erase the perception that it was our war, not a South Vietnamese war; and ineffectiveness of advice and resources devoted to aid the government of South Vietnam to make its case, improve its image, and enhance its performance on and off the battlefield. Responsibility for these failures did not rest solely—or even predominantly—with Joint US Public Affairs Office (JUSPAO). It was shared with the political and military leadership in Washington and in-country, with United States Information Agency (USIA) ambivalence about the merits of psychological warfare and its role therein, and with a military educational system that failed to inculcate in all ranks the fundamental tenets of counterinsurgency.

This evaluation in no way denigrates the dedication of the JUSPAO leadership and members. And, in all fairness, even programs of heightened impact would not have resulted in a different final outcome. Psychological warfare can enhance political, social, and military programs; but if those programs are seriously deficient, it cannot erase their faults.

The differences between the Philippine and Vietnamese experiences are dramatic.

• In both cases, the key objective was to win the support of the population. In Vietnam, that objective was not achieved. Admittedly, the Vietnamese government faced much more difficult challenges, explicit in a more numerous, better organized, more deeply imbedded, and more highly motivated opposition.

• Unlike in the Philippines, psywar personnel in Vietnam did not have direct and continuing access to the top political and military leadership or to the high-level planners. A number of first-class research projects, which were developed by social scientists, political psychologists, and cultural anthropologists, constituted guidelines for effective psychological initiatives. For the most part, these projects were simply filed.

• The Philippines' psychological programs were indigenous in content and means of communication. In Vietnam, impatience (to which short tours contributed) stimulated the Americans to do too much themselves. Not only was this conceptually wrong, it was also operationally ineffective because most of the Americans involved had no more appreciation of the political and cultural realities of Vietnam than we did of Lebanon in 1983.

Prospects for the Future

As the decade of the sixties ended and that of the seventies began, many observers believed that the United States would develop a comprehensive and prudent doctrine for psychological operations. In fact, some thought, the government would have no choice but to do so. Some even thought there would be extensive research to determine the most efficient and effective methods of applying psychological techniques and weapons.

Alas, it was not to be! The United States is still without the requisite national doctrine. Worse yet, there is less acceptance of the legitimacy of psychological warfare as a tool of statecraft by the State Department and the USIA than was the case in 1970. There has been scant modification of the long-held view that the conduct of psychological operations is an exclusively military task—appropriate in war, but deplorable in peace. The recognition has not dawned that such activities are fundamental to effective counterinsurgency efforts and must be primarily conducted by non-DOD agencies.

There have been some modest advances under the current administration. In 1984, a landmark National Security Defense Directive (NSDD) established international communications as a major instrument of national security policy and assigned specific tasks to executive departments and agencies. Only one section dealt with psychological operations—and then solely in a military context. It directed the revitalization of the armed forces' capabilities to conduct such activities in support of military operations in crisis and in war. Typically, the NSDD did not establish any effective interagency mechanism for orchestrating and coordinating the various international communication capabilities.

Using the NSDD as leverage, the secretary of defense directed the preparation of a Psychological Operations Master Plan (which he subsequently approved in toto). Its purpose was to remedy the deficiencies that had developed since Vietnam—deficiencies in force structure, operational concepts, planning, research, training, intelligence support, personnel programs, and understanding the potential of psychological operations as a military force multiplier. The plan included these important actions:

1. JCS development of a comprehensive joint doctrine to cover employment to strengthen deterrence, in crisis and in war, as the foundation of the revitalization effort;

327

2. augmentation of planning capability on major combatant command staffs;

3. functional separation of special and psychological operations at staff and organizational levels, formally recognizing that the latter is inherent in all military activity across the spectrum of conflict;

4. reindoctrination of the officer corps through appropriate instruction within the mainstream service school systems and including psychological operations in training exercises;

5. modernization of equipment and force structure; and

6. most significant of all, the phased development of a Joint Psychological Operations Center as the font for doctrine and conceptualization, for the direction of research and analysis, for planning support of the unified and specified commanders, and for assistance to the secretary and the chairman in developing defense positions on national psychological plans and campaigns.

It will take much hard work and continued command prodding to see these initiatives through to fruition. Absent an institutionalized mechanism at the national level, the psychological dimension of contingency planning is handled on an ad hoc, fragmented basis. DOD has repeatedly urged the establishment of a national psychological operations committee, which would, inter alia, do detailed planning for the marshaling and orchestration of all government and quasi-government communication capabilities in support of military contingency plans or external crises short of direct military involvement.

It is understood that the concept of the committee has been approved in principle, but has not been activated due to the strong reservations of the USIA. Conceivably, the congressionally mandated Board for Low-Intensity Conflict at National Security Council (NSC) level will provide impetus for the interagency psychological operations structure as an essential adjunct for sound planning.

Meanwhile, one Army unit (the 4th PSYOP Group) strives valiantly to plug the national gap. It has produced excellent how-to-do-it plans for focusing all relevant national capabilities in support of several countries under threat. The professionalism of the group is gaining recognition in key sections of the national security community. Both the group's planners and its colleagues on the ground can take much credit for the improvements that were wrought in the Salvadoran army.

Clearly, we are ill-prepared for the psychological dimension of the next major counterinsurgency that will engage the United States. The guidelines for enhancing performance are the major lessons gleaned from our past experience.

The first guideline—a truism meriting underscoring—is that the message communicated must be indigenous in content and in execution, leastwise within the confines of the supported country. The United States will have major tasks in ensuring political support for a beleaguered country on the international scene and here at home. In-country, the burden will fall on the shoulders of a few carefully chosen and highly qualified individuals who will operate in an advisory role, either formally or informally. The proportions of civilian and military advisors must be aligned to the structure of the host government; and while under the general supervision of the chiefs of the US Diplomatic Missions, they must work with and for their foreign colleagues.

Collectively, the US advisors must have a profound knowledge of the country—its culture, its customs, its political and social substructures, and the characteristics of its population. To ensure that depth of confidence, the group must be reinforced by—or be able to draw on the support of—the expertise of sociologists, clinical psychologists, and cultural anthropologists. They must, of course, be equally well informed about all aspects of the insurgency, its operating methodologies, its motivations, and its vulnerabilities. The challenge— a cross-cultural one—to those advisors will be to help the supported government shape and articulate value-based themes to which the population can relate. They must then ensure that all actions of the government and its personnel are consistent with—and reinforce—those themes.

A second, closely related guideline is to eliminate the mind-set that the practice of psychological warfare is the province of odd fellows steeped in the occult. To be sure, there are pressing requirements for full-time specialists and dedicated (albeit all-purpose) equipment. But these requirements deal

329

with only a portion of the psychological dimension in counter-insurgency: the detailed target research and analysis within both the native population and the insurgent structure; analysis of, and counters to, insurgent propaganda; and the offensive against the insurgents to confuse, confound, exacerbate vulnerabilities, and induce defections.

That sector aside, psychological operations (how badly we need other terminology!) do not have a separate compartment. They may at times be a part of public affairs, civic action, troop information, civil affairs, public diplomacy, humanitarian aid, or political action. As exemplified by the campaign against the Huks in the 1950s, everyone can and should aid in the psychological enhancement of all programs to gain and maintain the allegiance and support of the target audience: the population itself. The sine qua non for attainment of that objective are concrete programs that are sound in design, sound in content, and professionally implemented.

The complementary—usually crucial—task is to multiply the impact of those programs and underscore the relevance of those programs to national aspirations. Completion of this task will require unceasing exploitation of every medium of communication, from ensuring the proper deportment of soldiers in face-to-face contact with villagers to sending traveling shows throughout the countryside to presenting sophisticated TV programs.

Conclusion

My advocacy of the extraordinary importance of the psychological dimension at any level of conflict notwithstanding, I am not sanguine about our government's capacity to exploit this nonintrinsic instrument of national power (or, indeed, even to comprehend its potential for enhancing national security).

Most worrisome is the failure to recognize—let alone take measures to counter—psychological and political warfare waged against our policies with the intent to undermine them; here, the cost of inaction is real and heavy. The stark fact is that the American public, the media, and even the bureaucracy, are generally oblivious to the scope and sophistication of this genre.

South Africa commanded center stage in Western consciousness because there was a spectacularly successful (and
still ongoing) psychological warfare campaign. It was carefully
orchestrated, replete with misinformation, and aided and
abetted by our own gullible media. The Sandinista regime in
Nicaragua had similar success. Thanks almost exclusively to
the efficacy of its propaganda, there was more active support
among the American public for the Sandinista regime than for
the freedom fighters that our own government backed. Awareness that the integrity of our own political and decision-making
base is under threat should lead to actions that would expose
and neutralize our adversaries' efforts. Hopefully, such awareness will stimulate utilization of psychological and political
warfare initiatives to further the positive goals of American
foreign policy.

The Psychological Dimension of the Military Element of National Power

Maj James V. Keifer, USAF, Retired

> *War does not belong in the realm of arts and sciences; rather it is part of man's social existence. . . . Politics, moreover, is the womb in which war develops.*
>
> *If war is part of policy, policy will determine its character. As policy becomes more ambitious and vigorous, so will war.*
>
> —Carl von Clausewitz

The six-month Persian Gulf War, popularly called the One-Hundred-Hour War, recently took us through the continuum of the policy process. Economic sanctions were applied by a world community united with a degree of resolve seldom seen; but because these actions take time, and time favored Saddam Hussein, they were never given the opportunity to prove their effectiveness. Through diplomacy, essentially a government-to-government process of negotiation and compromise, the political leaders of the world tried to defuse the situation. Because both sides held nonnegotiable positions, with no room for compromise, the prospects for a diplomatic solution were nonexistent. Diplomacy did, however, play an instrumental role in uniting coalition forces. Was a peaceful resolution out of the picture? It evidently was!

While the political and economic elements of national power played out their hands, two others came to the forefront. These were the military and the psychological elements. These two elements, working together, seemed to constitute our best opportunity for a peaceful resolution; and after failing to achieve a peaceful solution, these two elements worked together to enable us to win, win quickly, and win with the least possible bloodshed and destruction.

How could the military instrument offer a peaceful solution? By itself, not at all. However, with the psychological element . . . ?

At the turn of the century, under Theodore Roosevelt's motto, Walk Softly and Carry a Big Stick, the United States practiced a policy called gunboat diplomacy. We built a reputation for "sending in the Marines" to achieve our national objectives. After a while, all we had to do was steam a ship to the coast; through intimidation, we achieved our objective.

Turning now to the recent Gulf conflict, we hoped to persuade Saddam Hussein to unconditionally withdraw from Kuwait through our military buildup in the Middle East. We felt that our recent excursions to Libya, Grenada, and Panama would certainly have impressed upon Saddam that we had the ability and resolve to settle the score militarily. And part of the psychological factor was Saddam's knowledge that his forces were equipped with hardware that was no match for our high-tech weapon systems. Another psychological factor was that, while Saddam didn't mind sending tens of thousands of Iraqis to martyrdom, he had no intention of becoming a martyr himself. The world community isolated Saddam and communicated unity on the issue through strategic psychological operations. Through theater PSYOP, we tried to convince him that it wasn't worth the cost to fight for a losing cause. Unfortunately, Saddam wasn't convinced; and neither were the Tikritis who supported him.

The psychological dimension played an important role in limiting the loss of lives and property. On the field of battle, we wanted to face an enemy who was unsure about his cause and capabilities but sure about his impending doom; an enemy who, even if unwilling to surrender, had little will to engage in combat. Through tactical PSYOP, we hoped to convey to the Iraqi military the futility of their position; we succeeded.

In the Middle East, the Andean Ridge, the Philippines, and throughout the world, the psychological dimension plays an important role in the achievement of national objectives. It serves the same purpose, and can be likened to, the aggressive marketing campaigns of Madison Avenue; that is, to affect the behavior of a target audience. Where Madison Avenue sells products and services, psychological operations sell ideas and ideology.

It's safe to say that all governments throughout the world make use of psychological tools during peacetime as well as during war. Activities are organized at two levels: international (we call these publicity, public diplomacy, and psychological operations) and in-country (we call these publicity, public relations, and public affairs).[1]

The use of any and every element of national power carries with it psychological effects.[2] To this end Murray Dyer wrote, "the highest levels of government now recognize that in any decision on national policy there are psychological factors that must be taken into account."[3] He cited the psychological dimension's cutting across departmental boundaries as the reason for an ongoing turf battle between the Department of Defense and the Department of State for control of the psychological instrument.[4]

The term *national power* describes the ability of a nation to influence other nations.[5] Deterrence, a fundamental national security objective of the United States for years, is essentially a psychological phenomenon.[6] Potential enemies perceive the cost of attack to be far greater than any possible gains. The assessment is based on the perception of this country's capabilities and resolve. Military psychological operations reinforce this perception by clarifying and repeating it. Edward N. Luttwak says it is an important message—one that must be communicated across the range of social and economic classes.[7]

Ronald D. McLaurin, author of *Military Propaganda: Psychological Warfare and Operations*, says that the reason for much of America's diplomacy and policy, both actions and words, is specifically to influence the behavior of other governments.[8] Martin Wright distinguishes diplomacy from propaganda:

> Diplomacy is the attempt to adjust conflicting interests by negotiation and compromise; propaganda is the attempt to sway the opinion that underlies and sustains the interests. Conversion therefore undercuts the task of compromise.[9]

How does the military element of national power support the psychological dimension? One fundamental foreign policy objective is to concentrate and optimize power. Other things equal, a nation that can back up its policies with coercive force holds a distinct advantage over less-capable nations. The projection of power is a central concept of international relations. Luttwak states that power, the influence of one state over others, may be measured in terms of the perception of the audience. To be effective (powerful), perceptions are precisely what foreign policy must affect.[10] The projection of US military power is often used as a force of political persuasion.[11] Such activities as port visits, bilateral exercises, firepower demonstrations, and special deployments help project power, as does the high-technology aura of US military hardware. Propaganda can be used effectively to reinforce and intensify these measures, but it must be carefully crafted to prevent counterproductive perceptions.

How, then, do psychological operations support strategic military objectives? The ability of nations to gather information, and to assess and interpret its relevance, isn't equal; what is perceived can be quite disparate from reality. With effects multiplied by differences in culture, experience, and expertise, the gap between perception and reality—misinterpretations—can lead to the misunderstanding of intentions, problems in resolving differences, and other dangerous situations.[12] Psychological operations bridge the gap.

If psychological operations are so important, why do they carry the stigma that seems to be attached? In the foreword to *Political Warfare and Psychological Operations: Rethinking the U.S. Approach*, Bradley C. Hosmer, president of the National Defense University, observed that "the negative connotations in the West of the word propaganda suggest we have treated political war as incompatible with democratic values and traditions." Frank R. Barnett, in the afterword to the same book, expressed also that "some would argue that the ethics of democracy preclude too strenuous a concern with propaganda." These statements unmasked American cultural bias toward the terms *propaganda* and *psychological operations*. Dr Carnes Lord, in "The Psychological Dimension in National Strategy," also noted this phenomenon:

> Manifest or latent in the attitudes of many Americans toward the practice of psychological-political warfare is a distaste for any sort of psychological manipulation or deception. The idea that psychological-political warfare is a black art that can be morally justified only under the most extreme circumstances is a derivative of

such attitudes. That such activities necessarily involve misrepresentation or deception is in any case far from the truth. But even assuming that some such element is inseparable from effective psychological-political operations, the moral calculus is by no means as clear as is frequently made out.[13]

The American attitude toward fair play reflects a disaffection for the use of any psychological manipulation. Gen Robert C. Kingston, commenting on "Political Strategies for Revolutionary War," by Richard H. Shultz, Jr., attributes this apprehension to past abuses of psychological tools by totalitarian regimes: "Many people in this country think that psychological operations equate to the Big Lie, suggestive of Goebbels and Hitler, and that we should not use it."[14] The late Gen Richard G. Stilwell, USA, Retired, provided comments on the institutional reluctance to consciously resort to perception for influencing activities. "The psychological weapon (and the political premises which govern its employment) have never been accepted as a permanent instrument of national security policy by the Executive Branch, the Congress, or the American people."[15] Cultural will, readiness, and the ability to conduct psychological operations have tended to surface only during wars. Ironically, it has been during wartime that the American public has become a primary target audience for national nonmilitary psychological operations.

Dr Lord credits the American media with helping to intensify cultural biases of the public through implications that underhanded objectives or purposes constitute the basis for such activities. He notes that claims of disinformation or public lies are often attached to media comments on psychological activities.

Dr Lord writes that the employment of psychological tools is further hampered by systemic weaknesses of the bureaucracy. From the top, he says, inadequate strategic planning and decision making prevent PSYOP from being considered, much less employed. The US diplomatic, military, and intelligence establishments distrust and resist any use of political-psychological warfare. Dr Lord charges that the State Department hasn't changed its diplomatic techniques in 200 years, ignoring the shrinking of the world brought about by dramatic advances in communications technology.[16]

337

The Department of Defense doesn't escape Dr Lord's criticism of institutional bias; he contends that, for reasons that are both similar to and different from those of State, military leaders have likewise tended to overlook the value of the psychological dimension, particularly outside tactical applications:

> The military services, in their preoccupation with technology, major weapon systems, and the big war, tend to neglect low-cost approaches to enhancing operational effectiveness, especially at the lower end of the conflict spectrum; and they tend to regard political-psychological warfare as someone else's business.[17]

Even the PSYOP community itself has struggled with its name for years, proposing such terms as *perception management, public diplomacy, political communication, influence operations,* and *international information.* In his article "By Any Other Name" in *Perspectives* (the magazine of the Psychological Operations Association), Col Thomas Taylor offered the title "Perception and Deception," suggesting the acronym, P&D.[18] Lt Gen Michael Carnes, director of the Joint Staff, urged a name change in his address to the Psychological Operations Association at the 1989 Worldwide Psychological Operations Conference, commenting that "you can't call soap 'dirt' and expect it to sell."

This struggle for a new name begs a couple of questions. General Carnes's "soap/dirt" analogy is faulty. First, the term *psychological operations* is amoral and the community is not responsible for bias against it. It would be more to the point that people think of "earth" or "topsoil" in terms of *dirt, filth,* or *grime.* And second, terms considered more acceptable—*politically* and *popularly*—fail to aptly describe the activity. A soldier in the battlefield making loudspeaker appeals to the enemy is not performing "political communications" or "international information."

Personally, I don't find the terms *psychological operations* or *propaganda* objectionable. What is needed is to market these words so that their pejorative connotations die out. Psychological operations help to save lives by using words, instead of bombs and bullets, to deter, or in the failure of deterrence, help win wars. In *The Weapon on the Wall,* Murray Dyer opines that the profession "is an honorable and important means, ranking with instruments of diplomacy, the military art, and economic measures."[19] One of his premises states that this discipline is a reflection of free will, a factor of public opinion. He's quick to point out, however, that this premise does not apply in a totalitarian state. Dyer says the way to overcome the institutionalized bias is through education. I agree: Education, coupled with advocacy, is the answer. The challenge is to encourage others to become disciples, too.

Notes

1. Ronald D. McLaurin, ed., "Organization and Personnel," in *Military Propaganda: Psychological Warfare and Operations* (New York: Praeger Publishers, 1982), 55.

2. Field Manual 33-5, *Psychological Operations* (Washington, D.C.: Government Printing Office, 1962), 78.

3. Murray Dyer, *The Weapon on the Wall* (1959; reprint, New York: Arno Press, 1979), 231.

4. Ibid., 69–73.

5. Martin Wright proposes that nations came into being with the decline of medieval Christendom when man's loyalty shifted to the state (king) and away from the estate (baron) and church (pope). He states the word *nation* came into being at the end of the French Revolution. A nationality was a collective people with a consciousness of historic identity expressed in a distinct language. Martin Wright, *Power Politics*, ed. Hedley Ball and Carsten Holbraad (Leicaster, U.K.: Leicaster University Press, 1978), 27.

6. Fred W. Walker, "Truth Is the Best Propaganda," *National Guard Magazine*, October 1987, 28.

7. Edward N. Luttwak, "Perceptions," in Ronald D. McLaurin, ed., *Military Propaganda*, 267.

8. Ronald D. McLaurin, ed., "Psychological Operations and National Security," in *Military Propaganda*, 2–3.

9. Wright, 89.

10. Luttwak, "Perceptions and the Political Utility of Armed Forces," in Ronald D. McLaurin, ed., *Military Propaganda*, 275.

11. Refer back to opening comments on gunboat diplomacy.

12. Luttwak, 267–68.

13. Dr Carnes Lord, "The Psychological Dimension in National Strategy," in Carnes Lord and Frank R. Barnett, ed., *Political Warfare and Psychological Operations: Rethinking the U.S. Approach* (Washington, D.C.: National Defense University Press, 1988), 22–23.

14. Robert C. Kingston, "Comments," in Carnes Lord and Frank R. Barnett, ed., *Political Warfare and Psychological Operations: Rethinking the U.S. Approach* (Washington, D.C.: National Defense University Press, 1988), 142.

15. The late Gen Richard G. Stilwell, USA, Retired, "Political/Psychological Dimensions of Counter-Insurgency," an essay of this publication.

16. Lord, 23–28.

17. Ibid., 27.

18. Thomas H. Taylor, "By Any Other Name," *Perspectives* 5 (Spring 1989): 9.

19. Dyer, 194.

PSYOP in *Desert Shield/ Desert Storm*

Col Frank L. Goldstein, USAF
Col Daniel W. Jacobowitz, USAF, Retired

Understanding the role of psychological operations in the Gulf War requires more than a glimpse of its contributions to military conflicts. To understand PSYOP, we must place it within the context of the times; specifically, in the sequence of events that led to Operations *Desert Shield* and *Desert Storm.*

When Saddam Hussein began to build his military strength along the Kuwaiti border, the United Nations (UN) began to react. Through a series of 13 resolutions, the UN demanded that Iraq end its aggression and pull out of Kuwait. Embargoes were established and ultimatums were delivered. Member states, including the US, Egypt, United Kingdom, France, Italy, Saudi Arabia, and Kuwait, deployed forces and/or equipment. Virtually the entire Western world—and a majority of Arab nations—opposed Iraq. Americans at home overwhelmingly came to the support of the president's actions (in contrast to the domestic situation during the Vietnam conflict).[1] The US government clearly articulated its policies and goals, having developed them through full, rapid, public debate, involving the executive branch, the Congress, and even the judiciary, which rejected various legal challenges to the evolving policies.

A new force, the international electronic news media, became a critical element in the successful execution of the war. In particular, coverage by the Cable News Network (CNN) demonstrated the power of international broadcasting and involved itself as a medium both to aid and to deter Iraqi PSYOP plans. In true democratic media tradition, the accumulation of factual reporting allowed knowledgeable audiences to correctly characterize Saddam's plans and behavior.

The Iraqi propaganda system is pyramidal, with the top spot occupied by Saddam Hussein. Control is maintained at the top, with many outlets, both overt and covert, at the bottom. Iraqi propaganda flows from the Ministry of Culture and Information (MCI) under the strict supervision of the Baath Party, the Revolutionary Command Council (RCC), and Saddam Hussein. Iraq's propaganda system was closely modeled after the Soviet system and is similar to that of most other totalitarian states.

An overriding aspect of Iraqi propaganda is the use of religious words and phrases. The techniques of Iraqi propaganda during the course of the crisis reflected the same style used by the Iraqis before the Gulf War. This campaign emphasized religious symbolism, Arab nationalism, and praise of Saddam Hussein. These themes reflected Iraqi culture and politics, demonstrating the fallacy of planning PSYOP based on the characteristics of the sender (rather than the nature of the receiver).

The Iraqis selected the following four objectives for their initial PSYOP campaign:

1. Rationalize their invasion of Kuwait.
2. Gain support of Arab masses.
3. Discourage nations from participating in the UN embargo.
4. Discourage or hinder military attacks on Iraq.

Four themes were generated from these objectives:

a. The revolutionary forces in Kuwait had asked for Iraqi help.

b. Iraq is the champion of oppressed Arabs.

c. The West is depriving the Iraqi people of food and medicine.

d. Iraq will withdraw from Kuwait after a short time.[2]

Later, a fifth objective—rationalizing the incorporation of Kuwait as a permanent province of Iraq—was established.

The MCI distributed the themes to the various media outlets for dissemination. This system worked very well, enabling the MCI/Baath Party to coordinate the campaigns, to react to impact indicators of current campaigns, and to initiate new campaigns on short notice. Some of the campaigns were selectively distributed to increase the stress impact on the target. Other campaigns used all media to ensure the widest possible distribution in the shortest amount of time.

Few limitations were placed on the type or content of Iraqi propaganda. At no time did missing or contrary facts interfere with Iraqi operations. If needed, documentation was simply manufactured. In the Iraqi PSYOP system, few restrictions are placed on efforts used to further an Iraqi goal. Furthermore, their campaigns did not have to follow any consistency, logic, or form; it was acceptable to criticize the Multinational Force's (MNF) bombing as being inaccurate on one day and, on the next day, to rest a theme on the destruction wrought by highly accurate MNF bombing. As far as Iraqi media were concerned, MNF bombing accuracy varied with the need to enhance the Iraqi propaganda campaign.

Saddam's use of PSYOP had varying results. Saddam used Scud attacks on Israel and Saudi Arabia as a political weapon to divert coalition attention and military effort from the main battlefield. The ever-present threat of chemical warheads was a major factor in his plan. According to Lt Gen Tom Kelly, operations chief for the Joint Chiefs of Staff, the Scuds were "of little military significance." Nevertheless, they produced emotional and psychological effects. Although Saddam induced fear among the Israelis and Saudis, and consternation among the coalition in general, his plan ultimately grounded on the rocks of military reality. However, Saddam's combination of useless military technology and a politically diverse targeting policy did serve to divert valuable coalition resources from other targets to the difficult task of Scud hunting.

Saddam's tactical purpose was achieved, but his strategic purpose was thwarted by dual coalition responses. Coalition tactical air efforts against the Scuds helped the Israelis maintain their policy of restraint, while the technological breakthrough of Patriot missiles calmed both Israeli and Saudi publics and further demonstrated coalition invincibility. Tactical military response, based on military competence and technological superiority, blunted a political weapon aimed at the heart of the coalition.

Iraqi efforts to use Western television as a relay for propaganda programs failed to convince important world audiences, although they may have been somewhat effective within the

Arab world. Among those efforts were Saddam's personal appearances with hostages, whom he attempted to characterize as "guests" of the Republic of Iraq. But Saddam's campaign was doomed by the perfectly natural and spontaneous negative reaction of a little British boy whom Saddam tried to convince of his good will. In this case the attempt to manipulate Western media completely backfired, as repetition of this scene throughout the Western world established public opinion of the true nature of the hostage situation. Further, it exposed Saddam's manipulative efforts and seriously eroded whatever credibility he retained.

Later Iraqi broadcasts were to fall flat as a result of Saddam's loss of credibility. For example, scenes of bomb damage at the otherwise unidentified building presented by the Iraqis as the Baby Milk Factory (with signs and workers' jacket logos conveniently in English) were quickly dismissed as crude propaganda by all except the most gullible or anti-Western. The nonsequitur of the signs and the behavior of the Iraqi media escort personnel (who limited reporter access within the site) conflicted with Iraqi claims. And when Western intelligence analysts demonstrated that the claimed bombing damage to cultural facilities in Iraq (the bombing of which is forbidden by the accepted laws of war) had in fact been faked by Iraqi wrecking crews, the remaining shreds of Iraqi television propaganda credibility were destroyed.[3]

Controversy still rages, particularly within the TV news industry, on the wisdom of continuing censored and pressured TV broadcasts from within the enemy capital. The TV broadcasts had both positive and negative tactical effects. Coalition operators were able to conclude from viewing scenes from Baghdad, or from the disappearance of live TV from Baghdad, that certain targets had been effectively struck.

In addition, the televised appearance of downed coalition pilots proved counterproductive for Iraq's objectives. Instead of convincing the coalition partners to acquiesce, a worldwide audience was appalled by the battered physical condition of the captives—and their orchestrated, mechanical admissions of guilt were painfully transparent. A backlash effect inspired greater support for the coalition.

A closer look at this one aborted Iraqi campaign will help clarify the propaganda process. The Iraqi goal of this propaganda campaign was to break the coalition. The Iraqi leadership concluded that if US public opinion could be marshaled against the war in the same way it was during the Vietnam conflict, the US would withdraw. Toward that end, the propagandists attempted to mimic some of the successful tactics used by the government of North Vietnam. One of the tactics employed was on-camera interviews of captured coalition pilots. The Iraqis thought that presenting these interviews would stimulate the US public to call for the withdrawal of US forces and the pursuit of peaceful options. Instead, outrage at Iraq's obvious violation of the Geneva convention resulted in even greater backlash against Saddam.

The highest echelons of the MCI/Baath Party made the decision to use captured allied pilots for propaganda purposes. The Iraqi News Agency, one of the principal media outlets to the West, managed the entire campaign. The plan to exploit allied pilots for propaganda included this strategy:

- capture several pilots,
- parade them through the streets,
- put them on television with prepared questions and answers,
- view the impact of the broadcasts through the Western media, and
- disseminate broadcasts to media and determine their impact.[4]

The initial problem that Iraq faced was how to disseminate their product; the coalition had damaged most of Iraq's television broadcasting capability early in the conflict. The Iraqis were forced to use a surviving low-power television transmitter, which they used to televise the parade of coalition pilots through the streets of Baghdad. But Iran was the only country that could receive their transmissions, and the Iranians did not share them with the world in a manner that suited the Iraqi purpose.

To circumvent this problem, the Iraqis videotaped interviews of the pilots and delivered them to various Western news agencies. The interviews of the pilots were choreographed with specific questions and answers.

Interviewer: What's your opinion of this aggression against Iraq?

Pilot: I think this war is crazy and should never have happened. I condemn this aggression against peaceful Iraq.

Interviewer: What do you think about this aggression against Iraq?

Pilot: I think our leaders and our people have wrongly attacked the peaceful people of Iraq.[5]

In the interviews the pilots told their families that they were being treated well. The MCI/Baath previewed and approved the interviews before releasing them to the media.

The response to the broadcast came very swiftly. All Western government, public, and media severely condemned the broadcasts and use of the downed pilots in this manner.[6] The worldwide condemnation was so overwhelming that the broadcasts, which had begun on 20 January, ended on 24 January. On 25 January, the MCI announced that the interviews of captured pilots would stop.

Another of Saddam's PSYOP efforts were the radio broadcasts to US troops in the field by "Baghdad Betty," reminiscent of those by Tokyo Rose during World War II. Intended to lower US troop morale, Betty's messages veered into the farcical as she warned American servicemen that their wives back home were sleeping with famous movie stars like Tom Cruise, Arnold Schwarzenegger, and Bart Simpson.

PSYOP, as used by Iraq, was intended to produce the greatest amount of negative stress within the coalition forces, the civilian population of those governments, and upon the civilian and political leadership of surrounding Arab states. As previously stated, the PSYOP efforts against the military forces of the coalition had minimal effects. However, the PSYOP effort aimed by Iraq at certain segments of the coalition countries and surrounding Arab nations was notable.

Saddam's propaganda effort as a stress-inducing instrument appears to have been as intense as the effort put forth by North Vietnam during the latter part of the Vietnam War. The

Iraqi PSYOP effort was clearly intended to pressure those nations supporting the coalition and those UN resolutions imposed on Iraq. Another Iraqi effort was to gather support for economic sanctions as an alternative to military action. Later, efforts were made against economic sanctions. Iraq portrayed the US, and specifically President George H. Bush, as the main opponent to Saddam. Some of the key themes used in these campaigns:

- The war was really about access to oil.
- The US was using the war as an excuse for imperialism.
- The US was propping up a corrupt government in Kuwait.

These attacks, targeted at the populations of Great Britain, Germany, France, Australia, Canada, the United States, and the Arab nations of the coalition, had varying degrees of success.

The Iraqi campaign against imperialism was a carbon copy of earlier Vietnam-era themes. Saddam spoke of the Gulf crisis as another Vietnam and painted a portrait of the US and coalition partners as embarking on another mistake. A great deal of effort was put into the following propaganda themes:

- The US was embarking on another dirty war.
- The war would be very bloody and would last a long time.
- The US would again be divided over the war.
- Again, the poor and the minorities are fighting a war for the rich and powerful.

Saddam Hussein's speeches were filled with images of certain doom, Vietnam-style—body bags returning home by the thousands, an enemy who could wait out the bombing to fight a bloody ground war, and an environment that would be as inhospitable as the jungle. He expected average citizens to recoil from the horror of such a war and spread more propaganda to dissuade the voters from supporting the coalition effort. A wedge between the Western nations would, in his mind, eventually result. The strategy is reflected in media writing around the world.

> Iraq is no Panama and Saddam Hussein no General Noriega. . . . The United States might be mired in the sands of the Arabian deserts just as they were bogged down in the paddy fields of Vietnam. (*New Straits Times*, 13 August 1990.)

> It was clear from the beginning that the Persian Gulf conflict would not only be resolved in the hot dunes of the desert, but also in the

quicksand of US public opinion. . . . While the public's reaction against the US involvement in Vietnam took years to materialize, it has developed very quickly [for the Middle East] . . . This is a fortunate development. (*La Jornada*, 22 October 1990.)

[Saddam] seeks to break Western solidarity and above all to put the Bush Administration in a corner. Time is in favor of Iraq now that US public opinion is gaining a clearer dimension of the high cost in human life of military action in faraway lands. (*La Nacion*, 22 August 1990.)

The military force building up against Saddam Hussein is such that a spark may be struck at any moment. The United States will be in danger of sinking into the sand, as it once did in the rice paddies of Vietnam. (*La Suisse*, 23 August 1990.)

There's the strongly-held belief that the [Vietnam] experience has rendered America incapable of marshalling public support to launch another war. (*Irish Independent*, 20 September 1990.)

While the Iraq propaganda effort certainly caused some confusion, and increased the stress of susceptible segments of the audience, neither public opinion nor the world media were moved as the Iraqi president had hoped. The public within the US and other coalition countries was determined to show support for their soldiers stationed in the Gulf. Neither slogans nor disrespect of military forces gained popularity during the war.

The failure of most Iraqi propaganda was in its credibility. While Saddam himself believed much of his propaganda, it appears that most of the world did not. His biggest successes were in the Arab world; and even in those countries, success was neither significant nor complete on any issue. The propaganda, in general, was far below the level of sophistication of effective audiences. Politically aware segments of the population, who would be inclined to be antiwar in general, were, if anything, turned off by the crude Iraqi attempts to manipulate their sincere beliefs. The clear articulation of goals, international support through the coalition and the UN, and the effective, low-casualty military action, rendered Iraqi propaganda ineffective.

Coalition Psychological Operations

In contrast to Saddam's efforts, coalition PSYOP efforts significantly affected Iraqi soldiers. First, the coalition efforts were of a more tactical nature (i.e., focused on convincing Iraqi troops to surrender). Second, the combining of PSYOP with the air and ground campaigns was intended to affect both soldiers and senior military leadership.

Tactical PSYOP in the Persian Gulf proved effective and efficient in terms of four principal sets of operations:[7]

- radio transmissions,
- loudspeaker broadcasts,
- leaflet disseminations, and
- enemy prisoners of war (EPW) team actions.

Operational success resulted from effective innovation in the field and from well-executed, time-sensitive planning. To a lesser extent, tactical operations proceeded from the deliberate planning model. Using the four identifiable sets of operations, PSYOP was implemented by various agents and organizations, ranging from theater-level cells to small, three-person teams serving in direct support of the forward combat units.[8]

The four operations showed different levels of effectiveness as to their impact on the Iraqi target audience. Table 1 lists the efforts and the relative effectiveness resulting from the four operations collectively.

Table 1

Exhibit I PSYOP Effort and Relative Effectiveness in the Persian Gulf

- - - - - - - - - - - - - Effort - - - - - - - - - - - - -

29 Million Leaflets Dropped In-theater
17 Hours Per Day of Radio Transmitting
19.5 Hours Per Day Aerial Broadcasting

- - - - - - - - - Relative Effectiveness - - - - - - - - -

73 Thousand Iraqis Reached through PSYOP
70% EPWs Report Messages Had Impact on Surrenders

Source: United States Special Operations Command (USSOCOM), "Post Operational Analysis: Iraqi Psychological Operations During Operations DESERT SHIELD/STORM," SOJ9, 1992.

The four operations focused on the intended Iraqi audience in concert with effective air and ground campaigns. However, because PSYOP had been slighted in the theater operational plan, this combination was not used at first. The result of this combined PSYOP, air, and ground effort was the influencing of an unexpectedly large number of Iraqi prisoners to surrender.

The four sets of operations—radio transmission, loudspeaker broadcasts, leaflet drops, and the actions taken by EPW teams—were conducted in January and February 1991.[9] The coalition initiated its tactical leaflet and radio activities in January 1991 to coincide with the start of the air campaign and terminated them in March and May, respectively. The loudspeaker and EPW actions began in February with the start of the ground campaign and ended in March and April.

An important precept at work in the radio and leaflet operations was reinforcement. Tactical PSYOP personnel announced to Iraqi ground units that a bombing was to occur at a specified time and place.[10] The next day, they announced that a bombing had indeed occurred as warned (only if the event had occurred as planned). The repeated cycles of announcement and execution helped persuade the Iraqis that the message and delivery means were credible and that surrender was a viable alternative to useless death.[11]

Radio Transmissions

Six broadcast platforms were established and used in the Persian Gulf theater of operations. Three were Volant Solo using EC-130 aircraft platforms (retransmissions made via modified USAF C-130 aircraft) and three were ground radio stations: "Voice of the Gulf," "Voice of America" (DOD-loaned equipment), and "Free Kuwaiti People." According to now declassified message sources, the radio stations were managed and run almost exclusively by US Central Command (USCENTCOM) as part of its theater operational plan. Thus, these stations were integrated into the overall combat-PSYOP effort relatively early. Programs consisted of pretaped messages broadcast continuously each day for about 17 hours. Messages conveyed such themes as the inevitability of defeat, Saddam's inappropriate leadership, and surrender appeals. A typical message transmitted over an established "surrender hotline": "YOUR DIVISION WILL BE BOMBED TOMORROW. ABANDON YOUR EQUIPMENT, LEAVE YOUR LOCATION NOW, AND SAVE YOURSELVES."

The relative effectiveness of these radio transmissions, in terms of audience exposure, was approximately 58 percent (versus zero for no transmissions). The degree of persuasiveness relative to no transmissions was estimated to be 46 percent. Impact on surrender reached about 34 percent.[12]

Such moderate values, according to some thinking, may be reflective of a culture whose primary representational systems (or language channels) seemed not to be auditory, but visual (as with leaflet drops) and kinesthetic (as with the humane actions of EPW teams).[13]

Loudspeaker Broadcasts

Typical PSYOP was accomplished in the theater of operations by two- or three-person loudspeaker teams in direct support of forward combat brigades. Teams consisted typically of one or two noncommissioned officers and an interpreter or communications specialist.

Loudspeaker teams broadcast messages that had been prepared and dubbed onto audiotapes and distributed to them and other teams by the 4th PSYOP Group by way of its product dissemination battalion. Occasionally, a team would ad-lib a broadcast if pressures of the moment demanded variation from the prepared script and if the language skill and initiative of the team so permitted.

Loudspeaker broadcasts generally produced moderate effectiveness in terms of audience exposure, persuasiveness, and impact on surrender. The rather uneven successes of the loudspeaker operations were found to be similar to those of radio transmissions, although some results, arguably positive, were obtained through the convincing appeals of enterprising loudspeaker teams. At least one team induced a captured Iraqi sergeant to make heartfelt appeals to his comrades across a berm using the very loudspeaker system that induced him to surrender.[14] Feedback from some EPWs indicated that, while "leaflets and radio showed us how to surrender, loudspeaker teams told us [exactly] where [to do it]."[15]

The following account by SSgt Edward Fivel, 9th Battalion, 4th Psychological Operations Group (POG), illustrates the operation of his loudspeaker team and its unusual effectiveness.

> We had to convince the company commander of our parent unit to let us PSYOP people try ousting those Iraqi soldiers from their underground bunker. We kept telling him there was nothing to lose by trying, especially since the pounding all morning by the 101st [Airborne Division] hadn't done anything.

> So the three of us gave it a try. We arranged for the Blackhawk [helicopter] to ride us to the site of the bunker and we began dropping surrender leaflets. Then we returned to base and found that nothing had happened—no surrenders, no movement, nothing.

> We tried something else. We hopped back on the Blackhawk and returned to the bunker area, this time intending to use our loudspeaker system and a pretaped message given by headquarters. We picked a spot on the ground about 800 meters from the bunker and started broadcasting. Nothing happened. I guess we couldn't be heard over the loud racket of our helicopter.

> We asked the pilot to land us on the ground not too far from the bunker. With what you might call serious reservations, he eventually landed us, feeling a little protected, I guess, by the three Apaches and one Blackhawk whopping above our heads and to our right. He took off immediately, saying he'd stay in contact by radio.

> The three of us were now on the ground. We were facing this enemy bunker that Intelligence says has 20 enemy soldiers who might be waiting to greet us. We picked up our loudspeaker equipment—the transmitter and the speaker—and ran about 200 meters closer to the bunker. We sat the equipment down and again started playing the cassette surrender tapes. Still no movement.

> Then our team leader decided to lift up the speaker. He lifted it and began carrying it even closer to the bunker. He toted that speaker exactly 50 more meters. I knew the distance because that's the length of the electrical cord he stretched out to the end, 50 meters. The team leader was very close to the enemy now.

> Then he suddenly stood straight up and pushed the speaker high over his head [like some kind of statue showing a big trophy to a crowd far away]. I told the other guy back with me watching all this, our communications man, to quit playing the taped message and go live, and to keep doing it. The guy had just gotten out of language school, so he could handle the Iraq language pretty well. He talked loudly through the speaker in four message sets.

That worked. A crackling voice came through our radio from the helicopter pilot who was still hanging up there with us. The pilot had spotted some movement. Then we saw Iraq soldiers begin climbing out of the bunker in front of our team leader. They were waving little white flags and carrying no weapons.

That was about it for us in that scene because our pilot said over the radio that he was landing to pick us up so he could refuel back at 101 division base. He was running real low on fuel. We didn't see how many Iraqis came out of that bunker, although we had counted up to about 20 of them before we took off.

When we got back to base, the three of us and our pilot began receiving weird congratulations. We weren't sure why. Seems that over 400 Iraqis eventually came out of the bunker, all without a fight.

For the next couple of days every helicopter the 101 had was flying back and forth carrying Iraqi soldiers from the bunker to a forward EPW camp. The camp must have been getting pretty crowded. The big thing we realized later was that these actions had kicked off the "Hail Mary" play in the ground war because they fixed the Iraqi units on line in front of us, letting our endrunners in the other division make their wide sweep around the main Iraqi force.

Leaflet Drops

Leaflets and other forms of print PSYOP proved especially effective. Of the targeted audience—300,000-plus Iraqi troops—approximately 98 percent of them read or were otherwise exposed to the 29 million leaflets dropped in the theater.[16] Many EPWs were found clutching leaflets in their hands or hiding them somewhere on their uniforms as they raised their arms to surrender.[17] An estimated 88 percent of the Iraqi forces were influenced by the leaflet drops as intended, and 77 percent were persuaded to quit the fight through the combination of combat-leaflet operations and credible tactical military threats and actions.[18]

Leaflets appeared to be effective in the Persian Gulf crisis. Their language was simple and straightforward. Appeals were visual for an audience that seemed to respond psychologically and emotionally to visual means. Whether the leaflets' effectiveness stemmed from psychological reasons or simply from the sheer volume of leaflets that descended on the Iraqis daily is open to interpretation. Weather conditions and low leaflet loss, combined with a generally effective theme, audience vulnerabilities, and effective coalition military action, resulted in an unprecedented success in inducing surrenders. The leaflets enabled many Iraqi soldiers, who were not highly motivated to support Saddam's war of conquest, to avoid sacrificing their lives in a doomed cause.

Still, some curious effects were attributed to at least some of the leaflet designs. These were effects that startled even the Arab partners who had supposedly provided input to the format.[19] PSYOP personnel later learned that the leaflets' effectiveness would likely have been enhanced had design consideration been given to the nonverbal, cultural cues that were read by the Iraqis as subtext.[20]

If this concept (subtext in language) escaped or seemed vague to designers and others at first, the application certainly became clear with each EPW whom they later interviewed during *Desert Storm* and similar situations.[21] Theater PSYOP personnel came to realize that they had missed an important dimension of cultural communication when designing the leaflets. In some cases, the PSYOP personnel were able to make adjustments and manage reprints; in others, they learned the cultural lessons after the cessation of hostilities.

An important illustration of this concept was the subtextual message transmitted by the color red. When red ink was included in leaflets, presumably for function or appeal, Iraqi soldiers typically hesitated or avoided approaching a leaflet sticking out from the sand nearby. They simply feared and disdained red as a signal for danger.[22]

Whenever Iraqis read a leaflet that showed a coalition soldier without a chin beard gesturing affably to an Iraqi, the reader became distrustful. To an Iraqi a chin beard signals a certain Muslim holiness and trust he could not find on a clean-shaven chin.[23]

When the Iraqis scrutinized a leaflet that showed Iraqi EPWs enjoying a bowl of fruit given by their captor, they were disappointed at seeing no bananas in the bowl. Bananas are a favored delicacy in Iraqi culture.

When an Iraqi soldier saw words encircled in the familiar thought bubble of Western comic strips, he became thoroughly confused. Unlike the English-speaking world and some other cultures, Iraqi culture does not link bubbles, words, and pictures.[24] However, this technique had been so internalized by the leaflet designers, that they were unaware that some cultures use other clues to indicate which cartoon character is speaking.

Overall, PSYOP in *Desert Storm* was among the most successful media-based PSYOP campaigns ever undertaken. Planners incorporated PSYOP into air operations from the outset and relied on leaflets to multiply the stress induced by unopposed bombing raids. *Desert Storm* coalition PSYOP proved the value of this combat adjunct at the tactical level and reaffirmed the overriding importance of credibility. Iraqi disdain for truthfulness contrasted vividly with Western punctiliousness in behavior and leaflet preparation. Coalition determination to stick to truthful overt PSYOP, relying on credible military performance to override minor cultural missteps in message preparation, proved successful in inducing thousands of early surrenders. The bottom line of coalition policy was met—casualties were reduced on both sides.

Notes

1. *Newsweek*, 25 February 1991.

2. USSOCOM, "Post Operational Analysis: Iraqi Psychological Operations During Operations DESERT SHIELD/STORM," SOJ9, 1992.

3. The US is not signatory to the 1954 Hague cultural property convention, but the US Army integrated the treaty interdictions in 1955. W. Hays Parks, fn 212, 60.

4. USSOCOM.

5. Ibid.

6. Such use of prisoners is forbidden by the laws of warfare. The failure of the Iraqi plan was partially engineered by such clever Yankee air pirates as Lt Jeffrey Zaun, who adapted the methods used by US airmen and sailors in the Vietnam War. Zaun augmented the battering of his face and exaggerated his inappropriate behavior to inform the world the Iraqis were maltreating him. What is more amazing is that the Iraqis would peddle these devastating tapes. The coalition pilot's actions were so effective that future conflicts may finally see the end of such illegal exploitation of prisoners.

7. USSOCOM.

8. Ibid.

9. Ibid.

10. Ibid.

11. Ibid.

12. Ibid.

13. Ibid.

14. Ibid.

15. Ibid.

16. Commander, 4th Psychological Operations Group, interview with author, 1991.

17. Ibid.

18. Ibid.

19. USSOCOM.

20. Ibid.

21. Ibid.

22. Ibid.

23. Ibid.

24. Ibid.

Epilogue

Lt Gen Samuel V. Wilson, USA, Retired

If past events are indeed prologue to the future, then understanding the early nature and scope of psychological operations, synthesizing the PSYOP lessons learned from historical national policy planning and from strategic, tactical, and operational PSYOP applications, can foreshadow even greater progress and success in accomplishing our national interests and objectives in political and military environments. *Psychological Operations: Principles and Case Studies* is the catalytic framework toward that end, since it reflects the foundation of PSYOP knowledge and wisdom.

It is critical that those involved with PSYOP—whether they are military commanders, political leaders, new PSYOP technicians, or experienced PSYOP practitioners and staff officers—become familiar with this book to better understand the importance that military PSYOP plays as a cost-effective, force-multiplier instrument of US military and political power. Such understanding is especially necessary today, given our military budgetary constraints, the existing nuclear, biological, and chemical threats, and our dynamic, complex, and worldwide arena of operations. Sun Tzu's strategy in 500 B.C. of subduing the enemy without engaging him is not only apropos but quite probably necessary for our very propagation.

The reader can learn much from parts II and III, which present various articles on national policy, PSYOP planning, and strategic and tactical PSYOP. Part III is especially significant since today much of the third world still uses Soviet PSYOP doctrine. In fact, those who would argue that the communist ideological threat has diminished with the demise of the Soviet Union, and that the US PSYOP focus should be abated, need to take heed of the various strategic and global resurgencies and the Vietnam legacy presented in part IV. One of the most important reasons for the communist "ideo-military" victory in Vietnam, for example, was that American strategic PSYOP had no political or moral acceptance as a necessary foundation of our national policy. We should not minimize the urgent need to increase American sensitivity to the psychological dimension of warfare, and these articles and case studies support that theme. The articles on our recent actions in Panama and in *Desert Storm* not only solidify the PSYOP requirement but also point out our weaknesses and future challenges.

These authors lead us to believe correctly that the PSYOP weapon system, if employed properly, must precede, accompany, and follow all military force employments while being closely coordinated with all agencies of government and while being systematically integrated with US national security policy and objectives throughout the spectrum of conflict and in peacetime. PSYOP is indeed a phenomenon in itself.

Contributors

Preston S. Abbott

George V. Allen, retired career foreign service officer, has served as ambassador to Iran, Yugoslavia, India, and Nepal, and as Greece's assistant secretary of state for Near East, South Asia, and African Affairs. Other positions he has held include director, US Information Agency; president, Tobacco Institute; and director, Foreign Service Institute.

MSgt Richard A. Blair, USAFR, a military intelligence analyst in the Reserve, is staff assistant to the senior military liaison and defense advisor, United States Information Agency, Washington, D.C. He was recalled to active duty during the Persian Gulf War to provide ongoing analysis of Iraqi psychological operations for Headquarters US Special Operations Command.

DeWitt S. Copp served as an Army Air Force Air Transport Command pilot in North Africa and the Middle East in World War II. Since the war, he has specialized as a book, film, and news writer in military aviation and foreign affairs. Copp has served on the staff of "Voice of America" as a writer-editor and with the United States Information Agency as policy officer on Soviet disinformation.

Col Benjamin F. Findley, Jr., USAFR, is a politico-military affairs officer with the Directorate of Psychological Operations and Civil Affairs, J-9, Headquarters US Special Operations Command. He holds a PhD degree in business. Colonel Findley has been a regular speaker at the USAF Special Operations School and a guest instructor at the Air University Center for Professional Development. He previously won the PSYOP Outstanding Instructor's Award and was the outstanding USAF Academy liaison officer for the Alabama-Northwest Florida

Region. Colonel Findley is a graduate of Air War College and is the author of several books.

Lloyd A. Free, president, Institute for International Social Research, School of Advanced International Studies, Johns Hopkins University. His former positions include editor, *Public Opinion Quarterly;* director, Foreign Broadcast Intelligence Service, during World War II; assistant military attaché, Switzerland; acting director, Office of International Information, State Department; advisor to Presidents Eisenhower, Kennedy, and Johnson; and advisor/consultant to Vice President Nelson Rockefeller.

Col Frank L. Goldstein, USAF, is dean, Education and Research, Air Command and Staff College, Air University. Formerly he served as deputy chief, Policy and Concepts Division, Directorate of Psychological Operations and Civil Affairs, J-9, Headquarters US Special Operations Command and as chief, Psychological Operations Branch, at the USAF Special Operations School. Colonel Goldstein earned a PhD degree in educational psychology and is a graduate of the USAF, British, NATO, and German PSYOP courses. He is a well-known writer and speaker among military audiences.

Col Joseph S. Gordon, USAR, professor of European studies at the Defense Intelligence College. He has taught history at Campbell College and at Duke University, where he earned a PhD degree in European history. Colonel Gordon's interest in psychological operations derived from three years as an intelligence analyst with the 4th Psychological Operations Group, Fort Bragg, North Carolina. He is a graduate of Army War College.

Col Daniel W. Jacobowitz, USAF, Retired, is assigned to strategic missile operations and the political-military field at Air University. Following his Titan II missile assignment, he had consecutive tours in Military Civic Action in Thailand. Colonel Jacobowitz has previously been on the faculties at Squadron Officer School and Air Command and Staff College and was operations officer of the Psychological Operations

Division, Combined Forces Command, Korea. After a tour in the State Department, he served as director of Psychological Operations in the Office of the Secretary of Defense and as director of the Center for Treaty Implementation, US Southern Command, Panama. He joined the Gulf War Air Power Survey team in 1991.

William F. Johnston is a member of the staff and faculty, National Interdepartmental Seminar, in the Office of the Chief of Psychological Warfare at the Foreign Service Institute. He has had several special warfare-type assignments, including command of a psychological warfare (psywar) battalion in US Army, Pacific. Johnston is a former chief of the Joint United States Public Affairs Planning Office, Vietnam.

Lt Col Philip P. Katz, USA, Retired, is a senior research scientist, American Institute for Research. He has been a psychological operations (PSYOP) instructor and lecturer at the Army Special Warfare Center, a PSYOP officer at the Department of Army, senior PSYOP officer for US Army, Pacific, and senior program manager for Development of Strategic Psychological Operations in Support of Field Activities, Vietnam. Colonel Katz developed a computerized PSYOP management information system for the Joint Chiefs of Staff and the Department of the Army. He is the author of several studies on psychological operations.

Maj James V. Keifer, USAF, Retired, is a psychological operations staff officer in the Directorate of Psychological Operations and Civil Affairs, J-9, Headquarters US Special Operations Command. He earned his master's degree in public administration from the University of Oklahoma and is a graduate of PSYOP courses instructed by the US Air Force and Army, the British army, and the North Atlantic Treaty Organization. Major Keifer's career has included experience in space operations, audiovisual media production and management, and public affairs.

Dr Carnes Lord earned PhD degrees from both Yale and Cornell Universities. He has been a distinguished fellow at the

National Defense University, an assistant to the vice president for National Security Affairs, and a foreign affairs officer with the Arms Control and Disarmament Agency. He has had consultant and director positions at the National Security Council and at the National Institute for Public Policy. Dr Lord's assignments at universities and colleges have been as assistant professor in the Department of Government and Foreign Affairs at the University of Virginia, instructor in the Department of Government at Dartmouth College, and lecturer in the Departments of Classics and Political Science at Yale University.

Ronald D. McLaurin is a research scientist, American Institute for Research. His previous positions include assistant for Africa, Office of the Assistant Secretary of Defense (International Security Affairs) and management assistant, Office of the Secretary of Defense. McLaurin is the author and contributor to studies on the analysis of foreign policy, international politics, and social and international conflict.

James Melnich is an analyst, Soviet Internal Affairs, Defense Intelligence Agency. He holds an MA degree from Harvard University. Melnich formerly served as director, Russian Émigré Center in Chicago.

Michael A. Morris is a fellow at the Canadian Institute of International Affairs. He formerly served on the faculty at Gallaudet University.

Laurence J. Orzell served as European branch chief, 6th Psychological Operations Battalion, 1st Special Operations Command, Fort Bragg, North Carolina. He taught European and American history at Marywood College and the University of Scranton. Orzell's publications on Polish history and immigrants have appeared in *Polish Review, Mid America,* and *Polish American Studies.*

Lt Col John Ozaki, USA, is a career military officer assigned to the Department of Army. He has served with the Army Security Agency, Headquarters 7th Army, Military Assistance Command, Vietnam, and in the infantry in Korea. Colonel

Ozaki is a graduate of Army Command and General Staff College.

Col Alfred H. Paddock, Jr., USA, Retired, served three combat tours with Special Forces units in Southeast Asia and held positions on the faculties of Army Command and General Staff College and Army War College. Other assignments include the Department of the Army Staff and the Secretary's Policy Planning Staff, Department of State. Colonel Paddock was commander of both the 6th Psychological Operations Battalion and the 4th Psychological Operations Group, and was director for Psychological Operations, Office of the Secretary of Defense. He earned his MA and PhD degrees from Duke University and is author of *U.S. Army Special Warfare: Its Origins.*

Gen Richard G. Stilwell (1917–91), USA, Retired, served in a number of key military and civilian positions throughout his professional life. He was commandant of cadets at the US Military Academy, commander of the 1st Armored Division and the 24th Army Corps, Vietnam, and chief of staff for General Westmoreland. As a civilian, the late General Stilwell served as under secretary of defense for policy and chairman of the DOD Security Review Commission.

Col Fred W. Walker (1942–90), USAF, Retired, was the first chief of the Psychological Operations Directorate, J-9, United States Special Operations Command. The late Colonel Walker was a pilot, parachutist, and past chief, PSYOP Division, Organization of the Joint Chiefs of Staff, Washington, D.C. An author and speaker on special operations throughout DOD, he was a graduate of the Armed Forces Staff College, Air War College, and the US Air Force Academy.

Col Dennis P. Walko, USA, master parachutist and licensed professional engineer, is chief of the Command Support and Coordination Division, Directorate of Psychological Operations and Civil Affairs, J-9, Headquarters US Special Operations Command. He has served in the Civil Affairs-PSYOP Exchange Office to the Colombian Army General Staff, as civic action coordinator to the El Salvador Combined General Staff, and as commander of the 1st Psychological Operations Battalion. Colonel Walko is a graduate of Command and General Staff College and the Inter-American Defense College.

Lt Gen Samuel V. Wilson, USA, Retired, is president, Blackburn College, Carlinville, Illinois. Before his retirement in August 1977, he had served in combat during World War II with the Office of Strategic Services, in the 1944 North Burma Campaign with "Merrill's Marauders," with the US Defense Attaché/Moscow, in Vietnam as associate director for field operations at the US Agency for International Development and, later, as US mission coordinator and minister-counselor for the US Embassy in Saigon. General Wilson has been Special Forces Group commander; assistant commandant of the Army's Special Warfare School; deputy assistant to the Secretary of Defense for Special Operations; deputy to the director of Central Intelligence; director, Defense Intelligence Agency; adjunct professor of political science, Hampden-Sydney College; chairman, Military Board of Virginia; chairman of the Secretary of Defense's Special Operations Policy Advisory Group; staff consultant to House and Senate Armed Services Committees; advisor on foreign technology, Los Alamos National Laboratory; and senior consultant at the Betac Corporation, Arlington, Virginia.

Lev Yudovich, born and educated in the USSR, emigrated in 1977 to the West. He is an international author of numerous articles on Soviet political and military subjects. He was a platoon and company commander in the Red Army during World War II and a former practicing defense attorney in Moscow. Yudovich has served on the staff of the US Army Russian Institute as professor of military and political science. He has written more than 200 articles for "Radio Liberty/Radio Free Europe" on Soviet political, military, and economic policy.

☆U.S. GOVERNMENT PRINTING OFFICE: 2003 – 633-004/84034